Practical Systems Biology

EXPERIMENTAL BIOLOGY REVIEWS

Environmental Stress and Gene Regulation
Sex Determination in Plants
Plant Carbohydrate Biochemistry
Programmed Cell Death in Animals and Plants
Biomechanics in Animal Behaviour
Cell and Molecular Biology of Wood Formation
Molecular Mechanisms of Metabolic Arrest
Environment and Animal Development: Genes, Life Histories and Plasticity
Brain Stem Cells
Endocrine Interactions of Insect Parasites and Pathogens
Vertebrate Biomechanics and Evolution
Osmoregulation and Drinking in Vertebrates
Host–Parasite Interactions
The Nuclear Envelope
The Carbon Balance of Forest Biomes
Comparative Genomics and Proteomics in Drug Discovery
The Eukaryotic Cell Cycle
Drosophila: A Toolbox for the Study of Neurodegenerative Disease

Practical Systems Biology

Edited by

CLAIRE GRIERSON
School of Biological Sciences, University of Bristol, Bristol, UK

ALISTAIR HETHERINGTON
School of Biological Sciences, University of Bristol, Bristol, UK

Taylor & Francis
Taylor & Francis Group

LONDON AND NEW YORK

Published by:

Taylor & Francis Group

In UK: 2 Park Square, Milton Park
       Abingdon, Oxon OX14 4RN

In US: 270 Madison Avenue,
       New York, NY 10016

Transferred to Digital Printing 2009

© 2008 by Taylor & Francis Group

First published 2008

ISBN: 9780415407809

A catalog record for this book is available from the British Library.

Library of Congress Cataloging-in-Publication Data

Practical systems biology/edited by Claire Grierson,
Alistair Hetherington.
   p. cm. – (Experimental biology reviews)
   Includes bibliographical references.

  ISBN 978-0-415-40780-9 (alk. paper)

  1. Computational biology. 2. Genomics. 3. Bioinformatics.
4. Proteomics. 5. Metabolism. I. Grierson, Claire. II.
Hetherington, A.M.
  QH324.2.P695 2008
  572.80285–dc22

2008003914

Editor: Elizabeth Owen
Editorial Assistant: Sarah E. Holland
Production Editor: Helen Powis
Indexed by: Shirley May
Typeset by: Keyword Group Ltd., Wallington, Surrey, UK

Taylor & Francis Group, an informa business

Visit our web site at http://www.garlandscience.com

Contents

Contributors   ix
Preface   xiii

1. Bioinformatic approaches to biological systems   1
   David R. Westhead, Iain W. Manfield and Christopher J. Needham
   1 Introduction   1
   2 Getting systems information from public data sources   2
   3 Using informatics methods to drive data generation   11
   4 Conclusion   15
   References   15

2. Cell sampling and global nucleic acid amplification   17
   Vigdis Nygaard and Eivind Hovig
   1 Introduction   17
   2 Cellular characterization and assaying gene expression   17
   3 Global analysis of the transcriptome – gene expression analysis at the RNA level   18
     3.1 Cell sampling considerations: Effect of sample properties on the transcriptome/
         transcript distribution to be assayed   20
   4 Approaches to reduce RNA requirements in microarray experiments   22
     4.1 Global mRNA amplification   23
   5 Concluding remarks   30
   Protocol 1: Linear RNA amplification   31
   References   34

3. Methods of proteome analysis: Challenges and opportunities   37
   Sarah R. Hart and Simon J. Gaskell
   1 Proteomics and the quantitative challenge   37
   2 Qualitative proteome analyses   38
     2.1 Common analytical workflows   38
     2.2 Protein recognition using conventional mass spectrometric data   39
     2.3 Protein recognition using tandem MS data   40
     2.4 Characterization of post-translational modifications   46
   3 Quantitative proteome analyses   48
     3.1 Relative quantification strategies   48
     3.2 Absolute quantification   52
   4 The kinetics of the proteome   53
     4.1 Protein turnover   53
     4.2 Kinetics of phosphorylation   53
   5 Concluding remarks and future challenges   54
   References   54

4. Vertical systems biology: From DNA to flux and back   65
    Annamaria Bevilacqua, Stephen J. Wilkinson, Richard Dimelow, Ettore
    Murabito, Samrina Rehman, Maria Nardelli, Karen van Eunen, Sergio Rossell,
    Frank J. Bruggeman, Nils Blüthgen, Dirk de Vos, Jildau Bouwman,
    Barbara M. Bakker and Hans V. Westerhoff
    1 Hierarchies in control: Living the dogma of molecular biology   65
    2 Hierarchies in regulation: Hierarchical Regulation Analysis   70
    3 Hierarchical Regulation Analysis: A practical approach   78
       3.1 An exemplary experimental analysis: Alcohol production by yeast as a function
            of time after nitrogen starvation   80
    4 Concluding remarks   88
    References   90

5. Using mathematical models to probe dynamic expression data   93
    Nick Monk
    1 Introduction   93
    2 The structure of interaction networks   94
    3 Mathematical models of interaction networks   95
    4 Linking state evolution models to expression data   95
    5 Differential equation models of cellular interaction networks   96
    6 Extracting information from expression data   98
       6.1 Example 1: Inference of a transcription factor expression profile, given
            time-course mRNA expression data for its target genes   99
       6.2 Example 2: Inference of the regulatory structure of a transcription network
            using protein expression data   101
    7 Using mathematical models to explore partial networks   103
    8 Concluding remarks   108
    References   108

6. Gene regulatory network models: A dynamic and integrative
    approach to development   113
    Elena R. Alvarez-Buylla, Enrique Balleza, Mariana Benítez,
    Carlos Espinosa-Soto and Pablo Padilla-Longoria
    1 Gene regulatory network models: Are they useful for understanding development? 114
    2 Mathematical tools for integrating biological processes at different time-space scales:
       Key for understanding pattern formation   114
    3 Dynamic gene regulatory network models: The Boolean case   118
    4 Epistasis and robustness: Two sides of the same coin for understanding
       developmental constraints?   121
    5 Dynamic GRN models for animal and plant modules   123
       5.1 Cell patterning   123
       5.2 Body plan development and evolution   126
    6 Genome level GRN models from microarray data: A challenge still ahead   127
    Appendix: Gene network inference methods: From raw microarray data to a GRN model 128
    References   135

7. Spatio-temporal dynamics of protein modification cascades   141
    Boris N. Kholodenko and Herbert M. Sauro
    1 Introduction   141

2   Computational modelling of growth factor signalling   142
3   Temporal dynamics of protein modification cascades   144
   3.1   The role of feedback   146
4   Spatial gradients of protein activities within a cell   150
   4.1   Waves of protein phosphorylation   151
5   Concluding remarks   152
References   153

8.   Intracellular signalling during bacterial chemotaxis   161
   Marcus J. Tindall, Philip K. Maini, Judy P. Armitage, Colin Singleton
   and Amy Mason
1   Bacterial chemotaxis   161
2   Intracellular signalling within bacterial chemotaxis   162
3   Mathematical modelling and bacterial chemotaxis   163
4   Developing a model of intracellular signalling   166
   4.1   Non-dimensionalization   170
   4.2   Parameterizing the model   171
   4.3   Model solutions and results   171
5   Summary and future work   172
References   172

9.   Modelling the mammalian heart   175
   Richard Clayton and Martyn Nash
1   Introduction   175
2   Physiological background   176
   2.1   Structure and function of the heart   176
   2.2   Cardiac cells   176
   2.3   The action potential   176
   2.4   Cardiac tissue and action potential propagation   177
   2.5 Ca^{2+}-mediated coupling of electrical and mechanical activity   177
   2.6   Tissue mechanics   177
   2.7   Mechano-electrical feedback   178
   2.8   Role of modelling   178
3   Cell models   179
   3.1   Electrical models   179
   3.2   Mechanical models   180
4   Cardiac tissue models   182
   4.1   Electrical models   182
   4.2   Mechanical models   183
5   Whole-organ models   184
   5.1   Anatomical models   184
   5.2   Whole-organ electrical models   185
   5.3   Whole-organ mechanics models   185
   5.4   Coupled electromechanical models   185
6   Numerical and computational issues   186
7   Two success stories   187
   7.1   Ion channel phenotypes of gene polymorphisms   187
   7.2   Models of fibrillation   188
8   Concluding remarks   189
References   189

10. Modelling root growth and development   195
    Eric M. Kramer, Xavier Draye and Malcolm J. Bennett
    1 Introduction   195
    2 Background to modelling root development   196
    3 Modelling hormone regulated root development   198
        3.1 A primer on computer models for auxin transport   198
        3.2 Model parameters   201
        3.3 Employing ISB to probe developmental processes in roots   202
    4 Towards a virtual root model   203
        4.1 Lessons from above   203
        4.2 The need to integrate many more signals   206
    References   207

Index   213

Contributors

Elena R. Alvarez-Buylla, Lab. Genética Molecular, Desarrollo y Evolución de Plantas, Departamento de Ecología Funcional, Instituto de Ecología, Universidad Nacional Autónoma de México, 3er Circuito Exterior Junto a Jardín Botánico, CU, Coyoacán, México

Judy P. Armitage, Department of Biochemistry, Microbiology Unit, University of Oxford, Oxford, UK and Oxford Centre for Integrative Systems Biology, Department of Biochemistry, University of Oxford, Oxford, UK

Barbara M. Bakker, Netherlands Institute for Systems Biology, Department of Molecular Cell Physiology, Vrije Universiteit, Amsterdam, The Netherlands

Enrique Balleza, Instituto de Ciencias Físicas, Universidad Nacional Autónoma de México, Campus Cuernavaca, Morelos, México

Malcolm J. Bennett, School of Biosciences and Centre for Plant Integrative Biology, University of Nottingham, Nottingham, UK

Mariana Benítez, Lab. Genética Molecular, Desarrollo y Evolución de Plantas, Departamento de Ecología Funcional, Instituto de Ecología, Universidad Nacional Autónoma de México, 3er Circuito Exterior Junto a Jardín Botánico, CU, Coyoacán, México

Annamaria Bevilacqua, Manchester Centre for Integrative Systems Biology, Manchester Interdisciplinary Biocentre, University of Manchester, UK

Nils Blüthgen, Manchester Centre for Integrative Systems Biology, Manchester Interdisciplinary Biocentre, University of Manchester, UK

Jildau Bouwman, Netherlands Institute for Systems Biology, Department of Molecular Cell Physiology, Vrije Universiteit, Amsterdam, The Netherlands

Frank J. Bruggeman, Netherlands Institute for Systems Biology, Department of Molecular Cell Physiology, Vrije Universiteit, Amsterdam, The Netherlands and Centre for Mathematics and Computer Science, Amsterdam, The Netherlands

Richard Clayton, Department of Computer Science, University of Sheffield, Sheffield, UK

Dirk de Vos, Netherlands Institute for Systems Biology, Department of Molecular Cell Physiology, Vrije Universiteit, Amsterdam, The Netherlands

Richard Dimelow, Manchester Centre for Integrative Systems Biology, Manchester Interdisciplinary Biocentre, University of Manchester, UK

Xavier Draye, Laboratory of Crop Physiology and Plant Breeding, Université Catholique de Louvain, Belgium

Carlos Espinosa-Soto, Lab. Genética Molecular, Desarrollo y Evolución de Plantas, Departamento de Ecología Funcional, Instituto de Ecología, Universidad Nacional Autónoma de México, 3er Circuito Exterior Junto a Jardín Botánico, CU, Coyoacán, México

Simon J. Gaskell, The Michael Barber Centre for Mass Spectrometry, Manchester Interdisciplinary Biocentre, Manchester, UK

Sarah R. Hart, The Michael Barber Centre for Mass Spectrometry, Manchester Interdisciplinary Biocentre, Manchester, UK

Eivind Hovig, Department of Tumour Biology, Institute for Cancer Research, The Norwegian Radium Hospital, Montebello, Norway

Boris N. Kholodenko, Department of Pathology, Anatomy and Cell Biology, Jefferson Medical College, Thomas Jefferson University, Philadelphia, Pennsylvania, USA

Eric M. Kramer, Bard College at Simon's Rock, Great Barrington, Massachusetts, USA and Centre for Plant Integrative Biology, University of Nottingham, Nottingham, UK

Philip K. Maini, Centre for Mathematical Biology, Mathematical Institute, University of Oxford, Oxford, UK and Oxford Centre for Integrative Systems Biology, Department of Biochemistry, University of Oxford, Oxford, UK

Iain W. Manfield, Institute of Integrative and Comparative Biology, University of Leeds, Leeds, UK

Amy Mason, Mathematical Institute, University of Oxford, Oxford, UK

Nick Monk, Division of Applied Mathematics, School of Mathematical Sciences, University of Nottingham, UK

Ettore Murabito, Doctoral Training Centre in Integrative Systems Biology from Molecules to Life, Manchester Interdisciplinary Biocentre, University of Manchester, UK

Maria Nardelli, Doctoral Training Centre in Integrative Systems Biology from Molecules to Life, Manchester Interdisciplinary Biocentre, University of Manchester, UK

Martyn Nash, Bioengineering Institute, The University of Auckland, Auckland, New Zealand

Christopher J. Needham, School of Computing, University of Leeds, Leeds, UK

Vigdis Nygaard, Department of Tumour Biology, Institute for Cancer Research, Norwegian Radium Hospital, Montebello, Norway

Pablo Padilla-Longoria, Instituto de Investigaciones en Matemáticas Aplicadas y Sistemas, Universidad Nacional Autónoma de México, Circuito Interior, CU, México

Samrina Rehman, Doctoral Training Centre in Integrative Systems Biology from Molecules to Life, Manchester Interdisciplinary Biocentre, University of Manchester, UK

Sergio Rossell, Department of Biotechnology, Delft University of Technology, Delft, The Netherlands

Herbert M. Sauro, Department of Bioengineering, University of Washington, Seattle, USA

Colin Singleton, Centre for Mathematical Biology, Mathematical Institute, University of Oxford, Oxford, UK

Marcus J. Tindall, Centre for Mathematical Biology, Mathematical Institute, University of Oxford, Oxford, UK

Karen van Eunen, Netherlands Institute for Systems Biology, Department of Molecular Cell Physiology, Vrije Universiteit, Amsterdam, The Netherlands

Hans V. Westerhoff, Netherlands Institute for Systems Biology, Department of Molecular Cell Physiology, Vrije Universiteit, Amsterdam, The Netherlands and Manchester Centre for Integrative Systems Biology, Manchester Interdisciplinary Biocentre, University of Manchester, UK and Doctoral Training Centre in Integrative Systems Biology from Molecules to Life, Manchester Interdisciplinary Biocentre, University of Manchester, UK

David R. Westhead, Institute of Molecular and Cellular Biology, University of Leeds, Leeds, UK

Stephen J. Wilkinson, Manchester Centre for Integrative Systems Biology, Manchester Interdisciplinary Biocentre, University of Manchester, UK

Preface

In 2003 the Nobel-prize winner Sir Paul Nurse gave a lecture at the University of Oxford entitled: The Great Ideas of Biology (The Romanes Lectures for 2003, Oxford University Press). In this lecture Sir Paul briefly discussed four established biological concepts: The Cell, The Gene, Evolution by Natural Selection, and Life as Chemistry. He ended by proposing that an emerging fifth idea, Biological Organization, might allow us to understand the cell as a logical and computational machine. Biological organization is the focus of interdisciplinary research, often known as integrative or systems biology, and carried out by life scientists, mathematicians, computer scientists, physicists, and engineers.

In this book we hope to provide an accessible first introduction to integrative and systems biology, inspiring and educating biologists, including life science undergraduate and postgraduate students, in the potential of systems and integrative approaches. We would also like to facilitate and encourage life science researchers to develop collaborations in this area. We have assembled a collection of examples, including technologies that are required for collecting good data, introductions to mathematical modelling, and examples of knowledge that can be gained by applying a range of interdisciplinary approaches to biological science questions.

This book grew out of a session at the SEB Annual Main Meeting in Canterbury in April 2006 and the New Phytologist meeting on Networks held in January of the same year. We would like to take this opportunity to thank the authors for finding time to write chapters. Thank you too to Professor Andrew Millar and Dr Ruth Bastow, who contributed to the organization of the meetings and made suggestions for this book, and to the speakers and session chairs at both meetings, who set the scene.

Bristol, December 2007   Claire Grierson & Alistair Hetherington

Bioinformatic approaches to biological systems

David R. Westhead, Iain W. Manfield and
Christopher J. Needham

1 Introduction

Bioinformatics is the science of handling and analysing large biological data sets. It saw a significant increase in importance in the early to middle 1990s as whole genome sequencing became a reality, and this has continued more recently as other methods of generation of high-throughput molecular data, such as transcriptomics, proteomics and metabonomics have emerged. Many systems biology projects are concerned with these data types, and rely on bioinformatics techniques and infrastructure.

In the context of data handling bioinformatics is concerned with the design of appropriate standards and databases for proper archiving. A well-known example of this is the MIAME (Minimum Information About a Microarray Experiment) initiative, an international project aiming to define the necessary information to enable a microarray experiment to be repeated or the data re-analysed. There are other similar initiatives in progress for other data types. A key extension of these ideas is the concept of *data integration*, so that diverse data types can be queried together, for instance linking transcriptomic data to genome sequences and annotations, or data related to the proteome and metabolome. This contributes to another key aspect: the need to provide convenient access to data and associated analysis tools for users with less well-developed computational skills, who wish to interrogate the data in order to ask biological questions. Although substantial progress has been made, these problems, and data integration in particular, remain far from solved. They are nevertheless very important for effective progress in systems biology.

Alongside the data, bioinformatics has developed a wide range of analysis methods. These range from simple BLAST searches used routinely by molecular biologists to find related gene or protein sequences, through an expanding number of more specialized tools for sequence analysis, to sophisticated statistical tools for the analysis of high-throughput data. The most basic and well-known application of bioinformatic methods is to annotate genome sequences. This involves analyses such as the prediction of gene structure and protein coding sequences, the prediction of protein function or structure by transfer of information from related or homologous sequences, the analysis of promoter sequence patterns and potential regulatory elements, and the identification of features such as potential membrane spanning segments, signal peptides and sites of post-translational modification. An important development alongside these prediction problems has been the creation of languages and controlled vocabularies for the description of gene function. The best known of these is the

Gene Ontology (GO), which aims to provide standard notation for aspects of gene function that are transferable (as far as possible) between species, and is structured so that relationships between terms are explicit, for instance recognizing that a term such as 'alcohol dehydrogenase' is a type of 'enzymatic function'.

Like wet laboratory biology, the bioinformatics field has begun to move from a focus on the functions of single genes (as in the annotation problem above) to increasingly consider the relationships between genes, as they operate together in pathways and networks. An example of this is the study of metabolic systems, where it is clear that the annotation of enzymatic functions of genes in a genome really only makes sense when these are considered in the context of the metabolic pathways into which the enzymes link, defining the metabolic capability of the organism. Another manifestation of this is through the GO project, where gene functions are described in terms of molecular function, which is typically a single gene property, but also in terms of the biological process(es) in which the gene participates (typically involving more than one gene arranged in a pathway, complex or network), and the cellular component(s) in which the gene functions. Thus bioinformatics has increasingly considered questions of *molecular systems biology.*

So, what is the contribution of bioinformatics to the new science of systems biology? The work described briefly above constitutes basic infrastructure that will be used by many molecular systems biology projects. In this chapter we consider more specifically how bioinformatics data analysis methods can contribute to systems biology projects, including methods to obtain systems information from the vast amount of public data available for many organisms. We will conclude, however, that while there is a vast amount of data available that contains useful information about systems, there is, in fact, nowhere near enough to give complete information about most systems. Systems biology projects will therefore need to generate a great deal of project-specific data, but more importantly bioinformatic or systems biology methods will begin to define what data is required. Thus the subject will move from a historical position of being driven by the generated data, to one where analysis methods will drive the generation of new data. Alongside these scientific objectives, bioinformatics will retain its role in archiving data to appropriate standards, and will contribute to similar methods for archiving systems biology models as well.

2 Getting systems information from public data sources

The term systems biology covers a wide variety of scientific investigations, but this chapter is restricted to studies of molecular systems where the high-throughput data typical of traditional studies in bioinformatics can make large contributions. Systems biology is characterized by the creation of mathematical models able to describe a system's operation and evolution in time, and predict its behaviour in varying circumstances.

Very often the systems concerned comprise components arranged in a network. In a metabolic system the components are metabolites and enzymes, where connections between metabolites are chemical reactions with enzyme catalysts. Multiple reactions connect to give pathways, which link together into a metabolic network. The system's state is specified by concentrations of metabolites and enzymes, and its evolution in time governed by fluxes through the reactions in the network. In a genetic regulatory network the components are genes and connections indicate regulatory influences; transcription factors (TFs) are typically given directed connections (arrows) to the genes they directly regulate. In protein signalling networks connections indicate the transduction of a signal, for instance a kinase might be connected to the proteins it phosphorylates. Hybrid networks are also possible, for instance linking signalling pathways to downstream effects on gene expression, or incorporating genetic regulation of the enzymes in a metabolic network. Indeed the integrated analysis of complete molecular systems comprising all of these network types is a key long-term goal of systems biology.

However, in many biological systems the list of molecular components is only partially known, and even the number of components may be uncertain. It is also often the case that the connectivity of the components in the network system is unknown. For instance it is rare to have a complete list of the target genes of a particular transcription factor, and it is often the case that experimental methods may show a connection between the expression of two genes without determining whether this connection represents a direct regulatory interaction, or whether it is mediated by some other, possibly unknown, component. Equally, it is often the case that signalling pathways are only partially known. Thus the first step in a systems biology project is often *system identification*, where the aim is to determine a sufficiently complete list of components and to determine their connections in an appropriate type of molecular network. Such a study usually precedes any effort to model the behaviour of a system. Generally, however, systems biology is an iterative process, and the comparison of the behaviour of a system model with experimental data can drive improvements in the model, including the discovery of new components or alterations to the proposed network.

Bioinformatic investigations can be very informative in systems identification. The following data types all carry information about biological systems, and for many organisms they are available in large quantity.

- Genome (and other) sequence data.
- Microarray (and other) measurements of the transcriptome.
- High-throughput data on protein interactions.
- Proteomic and metabonomic data.
- The scientific literature.

Genome data can be very valuable in system identification where the system concerned is well characterized in simpler organisms and known to be well conserved. This is often the case with metabolic systems, where metabolic networks, particularly the core of the network, are very well characterized in simple bacteria and yeasts, and conserved in basic form (although with some notable variations) across all kingdoms of life. The process of inference of the metabolic network for an organism is known as 'metabolic reconstruction'. Typically a first step in this process is simply genome annotation, where enzyme coding genes are identified by homology with enzymes in well-characterized organisms. The enzymes found to be coded in the genome are then mapped to standard or generic metabolic pathways, to give a first-pass estimate of the metabolic network. Typically such networks are imperfect, and reveal pathways for which only a subset of the necessary enzymes can be found in the genome sequence.

A better network can then be sought by examining the biochemical literature for evidence of species-specific metabolic information, including deviations from standard pathways; elimination of potentially false positive enzyme annotations revealed by isolation in the expected network; and by seeking potential novel genes coding missing essential activities. The research group of Bernhard Ø. Palsson at University of California, San Diego has done very impressive work on reconstructing metabolic systems for several organisms, and shown that mathematical models based on them can yield valuable biological information and knowledge. For example, they have shown that analysis of possible chemical flux distributions through the metabolic network can predict the growth behaviour of simple unicellular organisms such as *E. coli* and yeast under different nutrient environments and with genetic perturbations (e.g. gene knockouts; see for example Reed *et al.*, 2006).

Because of its conservation, the metabolic system is one of the easiest systems identification problems. Other systems can present more difficulties and the most appropriate investigations and data depend on details of the system. The experimental revolution that led to DNA (deoxyribonucleic acid) microarrays and the possibility of measuring the expression of every gene in a genome in a single experiment has created huge databases of microarray data. These data are often gathered by experimental groups with very particular purposes in mind, for instance working out which genes show differential expression under particular conditions, or in response to drug treatment. The product of such a study is usually a small number of interesting genes, and often the majority of the data is ignored. However, the data is valuable, particularly when it is added to a large database of other similar investigations, when it can be used to derive relationships in gene function way beyond those for which its generation was originally conceived.

To illustrate this use of microarray data, we consider the problem of defining a parts list for the system associated with a multi-protein complex/macromolecular complex. Proteins within the complex are required to act together, and their coding genes often show correlated patterns of expression. Other genes involved in the process but perhaps not as parts of the complex can also show correlated expression. In this example we used a large database of microarray data comprising hundreds of diverse experiments on the plant *Arabidopsis thaliana,* and examined the set of genes showing strong correlation (measured as the Pearson correlation coefficient, r) with the histone H4 gene, At5g59690. The analyses were done with our ACT software (Jen *et al.*, 2006). *Table 1* shows that the genes encoding histones comprising the nucleosome core octamer show strong correlation of expression, suggesting that their encoded proteins are likely to act together; indeed there are genes in the list shown for each of the expected proteins, that is histone H2A, H2B, H3 and H4. In addition, there are a number of genes with roles in DNA synthesis and chromosome organization which show strongly correlated expression.

However, many genes in *Arabidopsis* are present in gene families, arising from a series of genome duplication events. This is evident in the list in *Table 1* with multiple genes for histone H3 and H4. Duplication is a key evolutionary process, facilitating the development of biological complexity, as new copies of genes develop changed roles, regulation and expression. Examination of expression correlation lists for the ~22 000 genes in the database reveals that there are indeed other histone genes with expression patterns not strongly correlated with the group of histones under study.

Table 1. *The most correlated genes with an* Arabidopsis *Histone H4 gene At5g59690.*

r-value	AGI code	Annotation
1.00	At5g59690	Histone H4
0.85	At5g10390	Histone H3
0.84	At5g22880	Histone H2B, putative
0.84	At5g65360	Histone H3
0.83	At3g53730	Histone H4
0.81	At5g59870	Histone H2A
0.81	At2g40550	Expressed protein
0.79	At4g02060	Prolifera protein (PRL)
0.78	At2g42120	DNA polymerase δ small subunit-related
0.75	At1g67630	DNA polymerase α subunit
0.74	At2g16440	DNA replication licensing factor, putative
0.74	At1g09200	Histone H3
0.73	At2g07690	Minichromosome maintenance protein
0.73	At1g69770	Chromomethylase 3 (CMT3)
0.73	At5g41880	DNA primase small subunit family protein

r is the Pearson correlation coefficient

This is most clearly visualized by displaying on a scatter plot the correlation values for all genes with two query genes (*Figure 1*). In this case the query genes are the two histone H4 genes shown from *Table 1*. Expression of these two genes (represented by the two symbols at the top right of this figure) is strongly correlated. All other core histone genes are represented on this figure (see legend for details), showing that some of them are strongly correlated with the queries while other genes are poorly- or un-correlated with the queries. This suggests that transcript levels for these genes may be regulated differently, perhaps allowing the plant to respond more effectively to a range of environmental conditions. Histone H1 genes, encoding proteins with slightly differing roles from the core histone genes, also show a range of r-values for correlation of expression with the query genes. Such approaches allowing visualization of the behaviour of all (known) members of a gene family, against the background of all genes in the database, may help place different family members in the parts lists of different systems.

The use of microarray data in studying biological systems is often combined with the analysis of gene promoters to discover evidence of common transcriptional regulation. *Table 2* and *Figure 2* show an example of this type of analysis in action. In this case genes have been identified which have expression patterns which are strongly correlated over the same large database of microarray data with that of a plastocyanin gene encoding part of the photosynthetic apparatus (*Table 2*). The list shows genes that are known to be involved in photosynthesis, and also implicates some unannotated genes in this process. The r-values for correlation of expression of this set of genes with the query gene are high, suggesting strong similarity of expression patterns. Furthermore, analysis of the representation of GO terms attributed to this set of genes shows strong

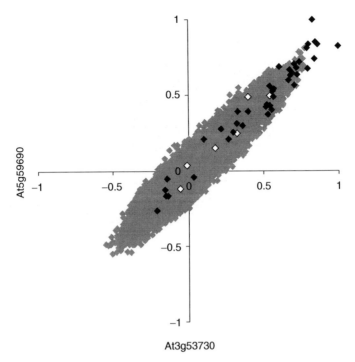

Figure 1. *Co-expression analysis of genes encoding sub-units of a multi-protein complex. The Pearson correlation values for ~22 000 genes versus two queries can be plotted against each other on a scatter plot. In this case two histone H4 genes have been used, represented by the two symbols at the top right of the figure. These and the values for other histone H2A, H2B, H3 and H4 genes are represented by black diamonds; histone H1 genes are represented by white diamonds and the remaining genes on the array are represented by grey diamonds.*

over-representation of the biological process and cellular component terms photosynthesis and photosystem. These statistical parameters lend confidence to the prediction that the unannotated genes are also likely to have some role in photosynthesis. *Figure 2* shows two DNA sequence motifs that are statistically over-represented in the promoters of the genes from *Table 2*. These were discovered using MotifSampler (Thijs *et al.*, 2002). Such motifs are putative *cis* regulatory elements that could represent transcription factor binding sites associated with common regulation of many of the genes on the list. In this case the motifs discovered are already known, representing the binding sites for bZIP and GATA transcription factors. However, these are large gene families in *Arabidopsis* and, although there may be strong candidates in the literature, there is very little information available, from chromatin immunoprecipitation (ChIP) experiments, on which genes are the *direct* targets of each protein. In contrast, in yeast, epitope-tagged strains have been created for most of the TFs in *Saccharomyces cerevisiae* and these have been used in ChIP-chip experiments (microarray analysis of immunoprecipitated chromatin) identifying many

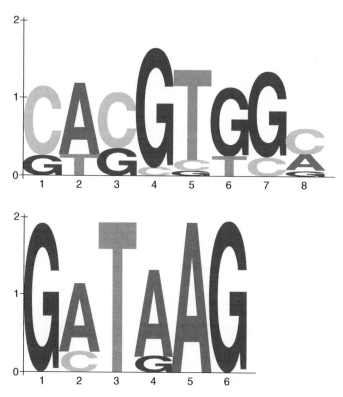

Figure 2. *Over-representation of sequence motifs in the promoters of the co-expressed genes from Table 2 supports the similar regulation of expression of these genes. These motif logos reflect the often degenerate nature of transcription factor binding sites. This analysis was performed using the MotifSampler website.*

combinatorial TF-promoter interactions. These results have then been combined with microarray co-expression analyses to identify gene modules and regulatory networks (Bar-Joseph *et al.*, 2003). The use of epitope-tagged proteins or protein-specific antibodies against plant transcription factors in ChIP experiments (perhaps using tiling arrays) will start to elucidate networks in plants, although *Arabidopsis* has ~10 times as many TF genes as *S. cerevisiae*. Developments in network prediction (discussed below) will help to reduce the numbers of likely regulators of each gene to numbers amenable to experimental testing using, for example, expression and ChIP analyses.

Figure 3 shows a final example of the use of co-expression analysis in systems identification. Co-expression analysis of a number of genes known to be expressed in pollen revealed diverse genes showing strong correlation of expression patterns with the query genes (data not shown). However, in this case there was no obvious theme to the annotations on the co-expression list and no significant over-representation of GO terms. However, employing information from another microarray database enabled corroboration of the list. To define the gene expression patterns characterizing

Table 2. *The most correlated genes with the* Arabidopsis *plastocyanin gene At1g20340.*

r-value	AGI code	Annotation
1.00	At1g20340	Plastocyanin
0.97	At2g06520	Hypothetical protein
0.96	At1g55670	Photosystem I subunit V precursor
0.96	At1g30380	Photosystem I subunit X precursor
0.96	At1g08380	Unknown protein
0.96	At3g21055	Expressed protein
0.95	At5g01530	Chlorophyll a/b-binding protein
0.95	At2g30570	Photosystem II reaction centre protein
0.95	At4g10340	Chlorophyll a/b-binding protein-like
0.95	At1g06680	Oxygen-evolving complex polypeptide
0.95	At4g12800	Photosystem I chain XI precursor
0.95	At2g26500	Unknown protein
0.95	At1g31330	Photosystem I subunit III precursor
0.95	At1g60950	Ferrodoxin precursor
0.95	At4g01150	Hypothetical protein

r is the Pearson correlation coefficient

specific cell types or in response to environmental signals, the AtGenExpress consortium has performed benchmark microarray analysis of many RNA (ribonucleic acid) samples, including specific dissected plant organs. This data has been combined with information from the description of each sample analysed by the Nottingham Arabidopsis Stock Centre microarrays group to produce the Genevestigator Metanalyzer (Zimmermann *et al.*, 2004) tool which gives an impression of the expression patterns for a set of user-input genes. Performing this analysis on a set of genes co-expressed with a pollen-specific calmodulin-binding protein (At2g43040) reveals that the co-expressed genes are also expressed in a pollen- or stamen-specific manner.

The type of analyses above can give information about systems without the need for anything but publicly available data. However, it is important to be aware of the potential limitations. The fact that microarray data measure very many genes means that most analyses contain some false positives (genes wrongly implicated in a system through high correlations that have occurred by chance instead of real biological phenomena). This is the well-known statistical effect of multiple testing: the chance of generating a false positive in any single test may be very low, but if you test many thousands of genes (22 000 in this case) then there will be some false positives. Further, depending on the nature of the data, the analysis will only be applicable to some molecular systems. In the case of the examples presented, the database contains diverse experiments on different plant tissues containing different cell types, grown in different environments using plants with differing genetic backgrounds. Correlation analysis over this data therefore really only reveals details of systems with a degree of universality, that is processes that occur in a coordinated manner in many cell types. The data is unlikely to reveal useful information on more specific processes occurring in limited cell types. Finally, in interpreting co-expression data it is always important

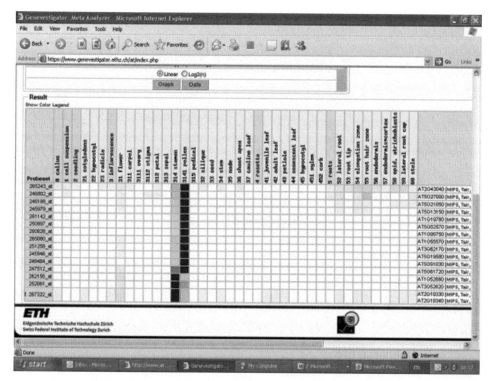

Figure 3. *Linking databases: Organ-specific expression of co-expressed genes. Databases of microarray results have been annotated for the source tissue of each experiment. The expression patterns of genes of interest can then be revealed across large datasets, such as are available in Arabidopsis, using, for example, tools available at the Genevestigator site. In this example, a set of co-expressed genes which showed no clear theme to their annotations show a clear organ-specific (pollen and stamens) expression pattern. This also indicates the value of MIAME compliance, allowing additional available information to be derived by pooling data to observe trends.*

to bear in mind that membership of the same regulatory system or network is far from the only explanation of co-expression. Genes from different systems can be co-expressed if, for instance, the systems occur in a similar set of cell types, or if they show similar temporal patterns of operation.

Because of its volume, microarray data has been very popular in the types of problems described above. However, more recently, high-throughput data on protein interactions from yeast 2-hybrid and tandem affinity purification has become available for a number of species and has been increasingly employed to solve similar problems. Typically this data defines a set of pairwise interactions between proteins, and its use in prediction of system membership is intuitive: proteins that are observed to interact are likely to be involved in similar systems. There are now a number of large databases of protein interactions available publicly. Since much high-throughput data is noisy, stronger evidence of systems membership can be obtained if predictions are corroborated in a number of independent data sources. For example, a prediction based on both gene co-expression and observed protein interactions is stronger then one based on a single source; this can be used to some extent to reduce the false positive predictions. Other bioinformatic approaches that have been employed in a similar way include the corroboration of interactions and co-expression in data from other species. Although this can only work if the process concerned is shared by the species concerned, the observation that interaction and co-expression are shared among orthologous genes in several species can constitute strong evidence of a functional association between those genes such as joint participation in a particular biological process.

So far we have considered only high-throughput experimental data as a source of information on biological systems. However, there is a wealth of useful information from the thousands of smaller scale experimental investigations that have been reported in the scientific literature over many years. Literature analysis is a bioinformatic problem. At its simplest level this is the provision of suitable search facilities enabling a user to define search terms and identify literature relevant to the system under study. Such a process is of course very much dependent on the skill and knowledge of the searcher, on the appropriate definition of search terms. It will be facilitated in future if those writing scientific papers can be persuaded to use controlled vocabularies and to provide structured abstracts that are more suited to computational searching. It should also be noted that a good deal of literature information is being included in bioinformatics databases, particularly protein interaction databases and sequence databases, by the process of expert manual curation. More sophisticated approaches to the literature include efforts to achieve automated information extraction. This extends simple keyword searching, which requires the user to read the retrieved literature and extract the important information manually; it involves the development of computer programs able to understand natural language and to extract facts, for instance that two proteins interact. This is a challenging task, because of the varied nature and structure of natural languages (there are many ways of saying that protein A interacts with protein B, using different words and grammatical structures, in constructions within the same sentence, between two sentences or even spanning paragraphs). Such approaches are starting to be useful, particularly to curators who currently spend large amounts of time including literature information in large databases.

A slightly different use of the literature to identify system components involves the concept of 'co-mentioning'. If two genes are consistently referred to together, for example they appear together in the same paper abstracts statistically more often than would be expected from the number of times they appear in the literature, then this can be used as evidence of a functional association between the genes, such as the membership of the same or similar processes. Although this might seem a slightly unusual idea, it is a relatively straightforward analysis to apply computationally to a large body of scientific literature, and it can reveal interesting information.

Beginning with microarray co-expression data, we have covered a number of bioinformatics information sources that can be used to predict functional associations between genes and hence to identify systems parts lists. In fact our listing has not been exhaustive, and to add to protein interactions, promoter sequence analysis and literature co-mentioning we could add the genomic context (genes located in the same chromosomal region can be functionally linked), the presence of gene fusions in some species, the phylogenetic profile (consistent gain or loss together in phylogenetic analysis indicates a functional link), and genetic association. All these can be predictive of functional association between genes. Systems identification using these diverse public sources of data is increasingly a probabilistic problem. Each source of information is more or less applicable to a particular system and is associated with its own set of problems and experimental noise. Research now focuses on how best to combine evidence from these diverse sources to give a reliable probability that the genes concerned are functionally associated. Two of the better established systems are STRING (von Mering *et al.*, 2005) and MAGIC (Troyanskaya *et al.*, 2003), each of which uses Bayesian statistics to combine evidence and predict functional association.

To conclude this section, we mention some inspiring research done by the group associated with Per Bork at EMBL Heidelberg (de Lichtenberg *et al.*, 2005; Jensen *et al.*, 2006), showing just how much interesting systems biology can be discovered by thoughtful analysis of available data. They studied protein complex formation during the cell cycle, first in yeast and then in other species. They combined two main data sources. DNA microarray data from a time series experiment was used to identify 600 yeast genes showing periodic expression and each was associated with a point of peak expression in the cell cycle. To complement this they gathered protein interaction data from yeast 2-hybrid experiments, pull down experiments and the literature, and filtered it to produce a set of high-confidence interactions. The strength of this combination of data is that the microarray data captures temporal information, but only identifies genes whose expression is regulated over the cell cycle, while the protein interaction data is static, but able to reveal information about complex components that are expressed constitutively. With 184 periodically expressed genes they were able to associate 116 interacting and constitutively expressed genes in cell cycle complexes. Some of the complexes were known, confirming previous knowledge, but many new components and modules were predicted. The work revealed that most complexes comprised both periodic and constitutively expressed genes, and suggested the hypothesis that 'just-in-time' assembly is the dominant control on cell cycle complex activity. In the second paper they revealed that this mechanism is shared by other eukaryotic species (the main cell cycle complexes are largely conserved), but that details of regulation, in particular the constitutive and dynamic components, show substantial differences between species. They presented evidence that loss of transcriptional control is often associated with new mechanisms of post-translational control. Thus the control mechanisms for the regulation of complex activity during the cell cycle have evolved.

3 Using informatics methods to drive data generation

The previous section provides some convincing evidence that analysis of available data can provide a great deal of information for systems biology. However, most biologists are interested in relatively specialized systems, and as we have already commented, depending on the system the methods above may or may not reveal useful information. It is also clear that while the methods can reveal 'parts lists', the amount of information revealed about network structure can be quite limited. In other words, the analyses can reveal sets of genes associated with processes, but they seldom show the details of connectivity such as the direct regulation of the expression of one gene by another, or which genes lie at the top of a signalling cascade and which are 'downstream'. Attempting to discover network structure from data is sometimes called a *reverse engineering* problem: an engineer would design the details of a system in order to achieve a desired behaviour, but we need the reverse of this, to work out what the system is simply by observing its behaviour. For most systems, reverse engineering will require the generation of data from experiments well planned for the problem at hand. Nevertheless, if a desired output of systems biology is a mathematical model, able to predict dynamic quantities such as gene/protein expression, then network structure knowledge is important. Only with this knowledge can simple and useful mathematical models of systems be built.

Network prediction has produced a myriad of different approaches in recent years and the literature can be challenging to follow. This is partly because terminology is difficult. The term 'network' can have a number of different meanings. We have already talked about metabolic networks, gene regulatory networks and signalling networks, which have meanings that are clear to most molecular biologists. But there are other more abstract networks, for instance protein interaction networks, where proteins are represented as nodes and connections are drawn between them if they 'interact'. Interaction can be anything from formation of a permanent molecular complex, to a transient physical interaction in a signalling pathway, to a more abstract functional interaction indicating involvement in a similar process. Another example is a co-expression network, where genes are connected if they show similar or correlated patterns of expression. With so many different network types it is unsurprising that the term 'network prediction' can have a number of meanings and lead to confusion. In this section, however, we focus on networks such as gene regulation and signalling, where connections between components indicate direct causal influences, for instance a transcription factor enhancing the expression of another gene, or a kinase phosphorylating another protein as a step in signal transduction. We are interested in data and analysis methods able to distinguish these direct interactions from other more indirect correlations.

Some simple example gene regulatory networks are shown in *Figure 4*, where transcription factors control the expression of other genes. Within such a network it is likely that the expression levels of all the genes involved will be correlated, and one might expect many of them to show up as co-expressed genes in an analysis like those presented with the ACT software earlier. The problem, in working out where to put

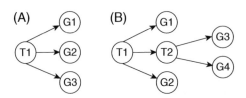

Figure 4. (A) A simple network motif where transcription factor T1 regulates the expression of genes G1,G2, and G3. (B) A more complex network where T2 is regulated by T1 and in turn regulates G3 and G4.

the connections in these figures, is to distinguish these general correlations from the direct regulation of one gene by another. The real difficulty in solving this problem is that even for a relatively small parts list there are very many possible networks of causal connection. For instance in the case of 10 components there are more than 10^{18} possible networks! If the right network is to be distinguished from all the other possibilities then a substantial amount of data is likely to be needed, and it will certainly help if this data is generated by carefully designed experiments aimed at network discovery.

Bayesian networks (Friedman, 2004) have become a very popular method for treating biomolecular network structures. The fundamental idea is that relationships between measurements on biomolecular components can be treated probabilistically, an assumption that fits well with the inherently stochastic nature of biological systems and measurements. Using the example of a gene regulatory network, the target of such an analysis is to determine the joint probability distribution relating gene expression levels. The probability distribution can be written, $p(G_1, G_2, ..., G_n)$, where $\{G_i\}$ represent the expression levels of a network of n genes. Bayesian networks are a way of simplifying such probability distributions that map in a very natural way to molecular networks. This is best illustrated by means of an example. Taking the network in *Figure 4B*, the joint probability distribution for expression levels is written $p(T_1, G_1, G_2, T_2, G_3, G_4)$ and involves six genes. But the network shows that, for instance, there is no direct causal relationship between the levels of G_1 and G_2: they are related simply because they are both controlled by T_1. In mathematical terms G_1 and G_2 are conditionally independent given T_1, $p(G_1, G_2|T_1) = p(G_1|T_1)P(G_2|T_1)$. Such conditional independence relationships allow simplification of the probability distribution, using the fact that each variable depends only on its parents in the network (those nodes with arrows pointing to it). In the case of *Figure 4B* the probability distribution simplifies to:

$$p(T_1, G_1, G_2, T_2, G_3, G_4) = p(T_1)p(T_2|T_1)\,p(G_1|T_1)\,p(G_2|T_1)\,p(G_3|T_2)\,p(G_4|T_2).$$

A method that can be employed to predict network structure from data is known as *structure learning*. Informally this can be thought of as finding the network structure that best explains the observed data. A variety of approaches are available to solve this problem. The simplest involve algorithms that search over many of the possible networks to find one that optimizes a score reflecting the quality of the network. A commonly used score function is the Bayesian information criterion, which incorporates how well the network fits the data, but includes a term to penalize the network complexity. More complex networks will almost always provide a better fit to the data, but increasingly good fits are often really only fitting to noise and unlikely to reflect biological reality. Methods therefore seek a compromise, to get a good fit with a reasonably simple network. A more thorough but computationally costly procedure is to use Markov Chain Monte Carlo methods to sample the probability distribution over networks and return a set of highly probable networks. This is an attractive approach, since biological data may not be sufficient to precisely define a single network.

The assumption of the above probabilistic approaches, when used in biology, is that the network connections that are most predictive of expression levels correspond to those that have a direct biological reason. In other words, although a gene may be correlated with a set of controlling transcription factors and also with other genes controlled by these transcription factors, the expression levels of the gene will be better predicted by the controlling factors than the other genes. While this is a reasonable assumption, there are probably cases where it breaks down. Equally, biological experiments often have missing data, and if a controlling component is missing then the method is likely to identify indirect but predictive interactions. Nevertheless, these methods have shown some notable successes on simple organisms such as yeast (Friedman, 2004).

It should be emphasized that the challenge in network structure prediction is the large number of possible networks, which poses computational challenges, but more importantly places enormous demands on the experimental data. In this field, experimental data is often noisy, and usually limited in volume. Even with DNA microarrays which measure expression across a whole genome, it is unusual to have even a few hundred relevant arrays, in a time series and other experimental arrangement, and the number available is often insufficient to determine networks. While there are ways of reducing the problem, for instance limiting the number of genes considered to those known to be involved, limiting the number of regulatory interactions to a biologically reasonable value, or focusing on known regulatory and signalling molecules and limited regulators to this set, this is really a fundamental problem. It is why theoretical methods like these need to start to drive data generation, so that the limited amount of data that can be practically generated is ideal for network structure determination.

A recent example of this is a flow cytometry study by Sachs and co-workers (Sachs *et al.*, 2005). They used Bayesian network methods similar to those above to reconstruct a human immune cell signalling network with high accuracy. The success of this study was due to the generation of a large volume of data in a way targeted to network prediction. Flow cytometry allows the simultaneous measurement of phosphorylated proteins and phospholipids in thousands of single cells. Measurements were made on several components known to be on the parts list for the pathway, with and without treatment with a selection of activators and inhibitors for some of the components. Activators and inhibitors are network *perturbations*, and they are very powerful in determining network structure. The reason for this is intuitive: any perturbation can only affect downstream components and therefore gives immediate network information. A study design using carefully chosen perturbations is likely to be a very efficient way of using scarce experimental resources to determine a network structure.

A different type of success story is that of Basso and co-workers (Basso *et al.*, 2005), who used microarray gene expression measurements to reconstruct the genetic regulatory network in B cells, employing a new statistical method related to Bayesian networks. They did not use perturbations, but instead relied on a large (~300 microarray) data set covering a large range of B-cell phenotypes. With this type of data it is not, in general, possible to discover the direction of connections between components, which would require either perturbation or time-series data. They generated a very large network and focused mostly on one hub (highly connected component) within it, the MYC transcription factor. MYC is predicted to regulate many genes and they verified a proportion of these experimentally. This study succeeded because a large volume of data was used, which covered a very large range of phenotypic conditions in a single cell type. It is rare to have such a data set; for most multicellular organisms microarrays are most often done on, at best, a single tissue, with a mixture of cell types.

Any signal reflecting network structure will be diluted if data are averaged over cell types where the signalling and regulation differ.

4 Conclusion

There are three principal challenges facing bioinformatics and its contribution to systems biology. The first is to leverage the existing public data sets to extract as much interesting biology as possible. There is so much data now, particularly on protein interactions and gene expression, and there is no doubt that it holds within it more interesting biology than has been extracted to date. To achieve this, more methods need to be developed, but a key challenge relates to the data itself. Although great progress is being made with database systems and standards, it remains the case that bioinformatics studies are still hampered by the need to spend significant time on the manual curation of appropriate data sets. The second challenge is to engage experimentalists more fully in the endeavour. Most bioinformatics groups need interesting biological systems to work on, and it is only with dialogue between experts in these systems that progress will be made. Finally, bioinformatics must respond to the challenge, and start driving the generation of data. Complete information about systems will only be discovered with extensive experimentation, informed in its design by theoretical methods such as those we have discussed.

References

Bar-Joseph, Z., Gerber, G.K., Lee, T.I. *et al.* (2003) Computational discovery of gene modules and regulatory networks. *Nat. Biotechnol.* **21**: 1337–1342.

Basso, K., Margolin, A.A., Stolovitzky, G., Klein, U., Dalla-Favera, R. & Califano, A. (2005) Reverse engineering of regulatory networks in human B cells. *Nat. Genet.* **37**: 382–390.

de Lichtenberg, U., Jensen, L.J., Brunak, S. & Bork, P. (2005) Dynamic complex formation during the yeast cell cycle. *Science* **307**: 724–727.

Friedman, N. (2004) Inferring cellular networks using probabilistic graphical models. *Science* **303**: 799–805.

Jen, C.H., Manfield, I.W., Michalopoulos, I., Pinney, J.W., Willats, W.G., Gilmartin, P.M. & Westhead, D.R. (2006) The *Arabidopsis* co-expression tool (ACT): a WWW-based tool and database for microarray-based gene expression analysis. *Plant J.* **46**: 336–348.

Jensen, L.J., Jensen, T.S., de Lichtenberg, U., Brunak, S. & Bork, P. (2006) Co-evolution of transcriptional and post-translational cell-cycle regulation. *Nature* **443**: 594–597.

Reed, J.L., Patel, T.R., Chen, K.H. *et al.* (2006) Systems approach to refining genome annotation. *Proc. Natl Acad. Sci. USA* **103**: 17480–17484.

Sachs, K., Perez, O., Pe'er, D., Lauffenburger, D.A. & Nolan, G.P. (2005) Causal protein-signaling networks derived from multiparameter single-cell data. *Science* **308**: 523–529.

Thijs, G., Marchal, K., Lescot, M., Rombauts, S., De Moor, B., Rouze, P. & Moreau, Y. (2002) A Gibbs sampling method to detect over-represented motifs in the upstream regions of co-expressed genes. *J. Comput. Biol.* **9**: 447–464.

Troyanskaya, O.G., Dolinski, K., Owen, A.B., Altman, R.B. & Botstein, D. (2003) A Bayesian framework for combining heterogeneous data sources for gene function prediction (in *Saccharomyces cerevisiae*). *Proc. Natl Acad. Sci. USA* 100: 8348–8353.

von Mering, C., Jensen, L.J., Snel, B., Hooper, S.D., Krupp, M., Foglierini, M., Jouffre, N., Huynen, M.A. & Bork, P. (2005) STRING: known and predicted protein-protein associations, integrated and transferred across organisms. *Nucleic Acids Res.* 33 Database Issue: D433–437.

Zimmermann, P., Hirsch-Hoffmann, M., Hennig, L. & Gruissem, W. (2004) GENEVESTIGATOR. *Arabidopsis* microarray database and analysis toolbox. *Plant Physiol.* 136: 2621–2632.

2

Cell sampling and global nucleic acid amplification

Vigdis Nygaard and Eivind Hovig

1 Introduction

Molecular biology has traditionally taken a reductionist approach to biological questions, focusing on a single gene, protein or reaction at a time, often using simple *in vitro* systems. This approach is hypothesis driven and assumes that the effects at the organism level can be reduced to causes at the gene level. Prior knowledge is required to some extent. In the past decade, however, the reductionist approach has been more or less replaced by global integrative approaches. This transition has been possible because of the recent availability of genomic sequence information and advances in high/medium throughout technologies and computer technology opening up for global approaches to assay cellular and molecular characteristics. Global assays can be used to test hypotheses, but are generally considered to rather generate novel ones. The most commonly applied global assays quantitatively or qualitatively query gene expression, which is the process of conveying genetic information from the DNA in the nucleus to the effecter molecules at the RNA or protein level. This knowledge is important for elucidating the complex relation between activation and deactivation of genes during physiological (e.g. cell cycle, differentiation, development, stimuli and response) as well as pathological processes (e.g. cause of disease, onset, consequences and progression). In the context of human health management, gene expression analysis may aid in the discovery of which genes may be potential targets for intervention in a therapeutic setting, and to elucidate how drugs and drug candidate work.

2 Cellular characterization and assaying gene expression

Multi-parallel methods to assay dynamic changes at all levels of gene expression have generated vast amounts of data in the exploration of cellular characteristics in relation to physiological and pathological states. Array-based technology, in particular, is a major contributor to the enormous amount of data accumulating and has been applied at all levels of gene expression analysis approaches, including genome (DNA), transcriptome (RNA) and proteome (protein) analysis. DNA-based comparative genomic hybridization on arrays (array-cgh) is a powerful approach to survey genetic instability, specifically copy number changes that potentially could have profound effects on gene expression levels (Lockwood *et al.*, 2006). Labelled genomic DNA from the cells of interest are co-hybridized against a reference DNA sample onto an array spotted with genomic sequences at a resolution that is determined by the size of the printed probes and the natural distance between these probe sequences located on the chromosome, for example one clone every 1 Mb or continuous coverage. The majority of global gene expression analysis, however, has concentrated on RNA-based methods involving microarray, serial analysis of gene expression (SAGE) and massive parallel sequencing signature (MPSS) technology. These powerful approaches query the RNA transcript abundance levels of thousands of genes simultaneously through either sequencing- or hybridization-based principles. Microarrays represent closed systems, analysing only the genes queried on the array-based platforms, while SAGE and MPSS are open systems with the potential to analyse all expressed genes in a sample/cell. For microarrays, single or dual samples can be hybridized and analysed on a variety of array types that are manufactured either commercially or in-house. The flow through of samples can be relatively high with the aid of automation at certain steps of the experimental procedure. SAGE and MPSS provide direct quantification of transcript abundances through transcript sampling of short sequence tags (10–22 bp), tag sequencing and identification. Typically the MPSS generates a larger tag library than SAGE yielding better transcriptome coverage. Requirement of high-throughput sequencing facilities or proprietary equipment affect their degree of utilization in the scientific community. This chapter is biased towards microarrays because they are the most common technology applied in global gene expression studies, although other approaches are also becoming more realistic for high-throughput experiments such as the qPCR-based OpenArray from Biotrove.

In contrast to DNA- and RNA-based analysis, direct quantification of protein levels at a global level is technically difficult at present, mostly because of the chemical complexity of proteins relative to DNA and RNA molecules (Bertone and Snyder, 2005; Maercker, 2005). The experimental conditions for measuring DNA copy numbers or RNA abundance are nearly the same for all nucleic acids; in contrast, the large range of optimal conditions that are specific for each protein molecule renders the establishment of high-throughput protein assays difficult. Development of high-throughput technologies in the field of proteomics has, however, led to a number of different array-based formats capable of screening a number of interactions including protein–antibody, protein–protein, protein–ligand, protein–drug, and enzyme–substrate. The capture array where ligand-binding reagents, usually antibodies, are used to target molecules in mixtures such as plasma or tissue extracts probably represents the most commonly applied protein array.

It is expected that by integrating DNA, RNA and protein profiles, along with gene annotation and gene ontology, functional correlation of genes and gene networks can be derived from these disparate data.

3 Global analysis of the transcriptome – gene expression analysis at the RNA level

As mentioned previously, transcriptome analysis is the most common approach to study the activity of genes, with the aim of elucidating biological processes governing the observed phenotypes of particular samples/cells. Analysis at this level implies measurements of the abundance or transcript copy numbers of specific RNA sequences to identify which genes (or elements) are transcribed, when and where (which tissues) genes are expressed, how many are expressed ubiquitously and to assess differences in gene expression between samples/tissues. Microarray experiments, like most other RNA-based methods, have generally been biased towards quantification of mRNA molecules transcribed from protein encoding genes, which, according to recent studies, only represent about 2% of the transcriptome in a mammalian study (Mattick, 2001). The transcript bias towards coding RNA is in the process of being altered because of the revelation of a vast number of non-coding RNAs with regulatory roles, for example microRNAs. Traditionally, however, the probes printed on the microarrays generally represent protein-encoding genes and the target labelling strategies typically select poly-(A) tailed transcripts for labelling from isolated total RNA prepared from the sample of interest. Polyadenylation is a common feature of mature mRNA molecules. The estimated number of protein encoding genes in the human genome lies between 20 000 and 25 000 unique genes. However, because of the existence of numerous splice alternatives of protein encoding transcripts, the use of alternative transcription start sites and/or alternative polyadenylation sites, one protein encoding gene may be represented by several unique probes on an array. It is also worth noting that the fraction of transcripts that is labelled includes not only protein encoding transcripts, but also the fraction of expressed non-coding RNA harbouring poly-(A) tails, in addition to contamination of unprocessed mRNA transcripts. During general labelling procedures, size selection of labelled targets filters out a portion of non-coding RNA transcripts as a result of their short lengths.

If we consider only the protein encoding fraction of the transcriptome, how many genes are expressed in a typical mammalian cell? Early studies estimated this figure to be 12 000 genes per cell type (Hastie and Bishop, 1976), which is in agreement with more recent findings based on analysed SAGE libraries (Velculescu et al., 1999). Hence it would be fair to say that, if a global gene expression analysis of a pure cell population detects a reasonable number of gene-specific transcripts with respect to the estimated reference figure of 12 000 in the experimental system, then the transcriptome complexity of the sample has been preserved throughout the procedure. However, the observed complexity is a function of the total frequency/abundance of the gene specific transcript assayed and the sensitivity of the technology. Hence, the next question that arises is: What is the frequency distribution of the expressed transcripts in a typical mammalian cell? The mRNA of a typical somatic cell may be divided into three frequency classes, high-, medium- and low-expressing genes (Bishop et al., 1974). Although the true frequency is unknown, results from expression assays indicate that the distribution is likely to differ between cell types (Reverter et al., 2005). Hybridization kinetic studies have indicated that the distribution of these three abundance classes is one of several possible categorizations (Quinlan et al., 1978). It has also been shown that the number of abundance classes within a type of cell may vary between unstimulated and stimulated groups, transcription being heavily induced in stimulated cells (Quinlan et al., 1978). Regardless of the discrete or continuous copy number distribution, it is apparent that transcript abundances do not follow a normal distribution, but are skewed, having a heavy tail of low-expressing genes. According to SAGE studies, more than 83% of the collective transcripts were present at levels as low as one copy per cell (Velculescu et al., 1999). The seemingly simple, but intriguing question regarding general features of the transcriptome is: What is the total number of transcripts per cell? Here we do not have much insight, but keep relying on a 30-year-old reference from 1974 (Bishop et al.) which estimated 300 000 mRNA transcripts per cell. This figure is essential with respect to the probability of sampling of transcripts for accurate quantification through direct or indirect procedures, and it is therefore rather intriguing that the persistent reference from 1974 has only recently been challenged (Nygaard *et al.*, manuscript in preparation). The authors suggest a re-evaluation of this number based on their microarray data, which indicate that the total value exceeds 1 000 000 transcripts per cell.

3.1 *Cell sampling considerations: Effect of sample properties on the transcriptome/transcript distribution to be assayed*

3.1.1 Cellular composition

Sample selection is a crucial step in the experimental procedure, as it defines the features of a sample to be analysed. A thorough consideration and insight into the sample features may alleviate some of the complexity in the downstream analysis. An important aspect of the gene expression measurements obtained by analytical tools is that the values represent the average transcript levels per gene from all the cells in the selected sample. Whether the sample is relatively homogenous or heterogeneous, with respect to cell type composition, is therefore a relevant consideration when addressing the results. The more pure the cell population is, the easier it is to assign distinctive gene expression patterns to specific cells, not confounding the results with contaminating cells. This implies that if the samples are heterogeneous in nature, then a larger set of samples may be necessary in order to have statistically significant data.

3.1.2 Time of harvest

Furthermore, in the sampling process the time of harvest may be an important consideration in order to detect relevant information about the response or progression of originating events. The transcriptome is highly dynamic and changes rapidly in response to cellular events, such as cell-cycle state, or perturbations caused by inducing stimuli (Darzacq *et al.*, 2005; Ueda *et al.*, 2004). Hence, the handling of the biological material should strive towards minimal disturbance that could induce changes in gene expression.

3.1.3 Sampling strategies

The biological material to be analysed and the underlying aim of the investigation should in principle dictate how we practically collect the sample for gene expression analysis, although there are a number of issues impacting the investigator's decision which are unrelated to the overall biological aim. These include feasibility and availability of sampling/collection technology, and the resulting RNA quality and yield.

The sample can be collected at several levels of refinement, from the most applied level, fresh-frozen bulk/whole tissue or sampling over a wide range of cell numbers, to tissue sectioning or cell enrichment, and even down to microdissection or single cell picking. In a clinical setting, patient samples come in a variety of sizes and cellular compositions from both invasive and non-invasive procedures. Certainly small non-invasive clinical specimens relieve sampling and storage procedures, but only yield a few micrograms of total RNA or typically even less. Expression profiling of bulk tissue probably represents the majority of published microarray studies. However, the question is whether it is possible to decipher the complex expression patterns from cells of interest, considering the heterogeneity of the cell types present in bulk tissue. Studies have shown that gene expression profiling of bulk tissues yields the profile of the dominating cell type, while minor cell types are washed out (Szaniszlo *et al.*, 2004). Hence, profiling bulk tissue has it limitations. Advances in technology designed for selective collection of specialized cells, coupled with an increased range of input requirements to microarray experiments may clearly alter the application status of bulk tissue samples in the future. However, with respect to enrichment steps, there are issues of additional handling and reduced sample sizes. Enrichment methods, such as fluorescence-activated cell sorting (FACS) or immuno-magnetic purification, allow selection of live cells for further analysis, while laser capture microdissection (LCM) relies on fixed and stained tissue. All of these procedures could possibly influence the final quality of the extracted RNA. The possibility of selecting a few, or even single, cells, sounds intuitively like an optimal strategy that yields maximum specificity regarding the gene expression profile for a particular cell type, but technically challenging issues and, as we shall discuss later, analytical issues arise that burden the biological interpretation of the data results.

3.1.4 Sample/RNA quality

Having selected and defined the sample to be analysed, RNA needs to be extracted and purified, although some strategies omit this step by targeting the RNA molecules directly in the cellular lysate. Disaggregation of tissue and preparation of cells requires immediate transfer to a stable environment for RNA, because RNA is extremely delicate once removed from the cell, and is prone to degradation if not working in an RNase-free environment. This can be done by immediate processing of the sample with an RNase-free lysis buffer, flash freezing in liquid nitrogen or storing the sample in an RNA stabilizing solution such as RNAlater® (Ambion). Cellular suspensions are typically washed in phosphate-buffered saline (PBS) before freezing or lysis. The method applied to the isolation of total RNA may affect the quality and quantity of the RNA yield. There are a number of commercial kits available for total RNA isolation, of which several are suitable for handling small samples, such as NanoPrep from Stratagene. Quality controls using spectrophotometers or lab-on-a-chip systems (Bioanalyzer, Agilent Technologies Inc.) are helpful in assessing the quality/ degree of degradation in an RNA sample. Degraded or contaminated RNA samples tend to cause deterioration of downstream experimental quality.

With respect to the RNA extraction step, we generally assume that the isolated RNA represents all species of mRNAs in the same proportions as in the original cells, although this is not necessarily true and difficult to verify.

3.1.5 Sample size/amount of RNA available

The sampling process and final RNA extraction yield dictate our options with respect to microarray experiments. The introductory microarray technology paper stated that the amount of mRNA used for target preparation was 5 μg (Schena *et al.*, 1995). In terms of total RNA, this figure converts to 165–500 μg assuming 1–3% mRNA content and not all investigators could provide such substantial amounts of RNA from their cells of interest. Clearly there are a number of sampling procedures mentioned above that are far from producing the amount of RNA necessary. Strategies to circumvent this issue are the subject of the next section.

4 Approaches to reduce RNA requirements in microarray experiments

Overcoming the hurdle of large material requirements for a single microarray hybridization experiment has received particular attention in the microarray community because of the restriction of a number of samples, such as clinically important biopsies and fine-needle aspirates, in addition to attempts to reduce complexity of samples by procuring small enriched populations or laser-captured material. However, it has also been necessary to address the RNA amounts required for each analysis in the alternative high-throughput techniques SAGE and MPSS.

One option to address the lack of sufficient input RNA is to pool the RNA from individual samples. Although it may be practical and reduces costs, discussions of validity are prevalent, typically within the microarray community. When dealing with patient samples, for example, the assumption is that the variation of a gene among different individuals is approximately normally distributed. In practice, the common observation is that there are more outliers than in a normal distribution. Hence, a consequence of pooling samples is that genes that are identified as differentially expressed may turn out to be extremes in only one individual. The gene expression averaging across pooled samples reduces the chance to observe sample variability and weak gene expressions may be washed out, but substantive differences may be easier to detect (Dobbin *et al.*, 2003). Hence, there is a trade-off regarding the possible complex expression, which requires some consideration on the part of the investigator, that is dependent on the estimated variance between the samples to be pooled. Effects and guidelines for RNA pooling strategies have been presented in the light of microarray experimental designs (Kendziorski *et al.*, 2005; Zhang and Gant, 2005). Another solution is to culture the limited cells of interest to expand the population size. However, the drawback of using short-term cultures is that cells are separated from their natural microenvironment and changes in gene expression as a result of handling could potentially confound biological findings.

Efforts to substantially reduce the large RNA requirements have focused on two main strategies, signal amplification and sample amplification. The first strategy involves improving and optimizing the target labelling reaction in order to increase the number of signal molecules per transcript. Augmenting the number of signals can be achieved by technologies such as dendrimer (Stears *et al.*, 2000) or tyramide signal amplification (TSA; Karsten *et al.*, 2002), in addition to enhancing and optimizing target preparation based on standard protocols. Currently, the minimum number of cells required by commercial products offering either optimized procedures or signal amplification is $2.5–7 \times 10^4$ cells. This number still exceeds by far the quantity of cells found in certain samples (fine needle aspirates, FNA) or small purified cell populations (LCM or immunomagnetically selected) and requires alternative strategies to solve the problem of RNA insufficiency.

Development of sample amplification procedures has been relatively successful in challenging the obstacle of insufficient material. The purpose of sample amplification, or global mRNA amplification, is to increase the number of gene transcripts to sufficient quantities for the microarray experimental procedure. The up-scaling of mRNA species can be performed either linearly, using T7-based *in vitro* transcription, or exponentially by PCR-based strategies or a combination of both.

4.1 *Global mRNA amplification*

Before the microarray era, Van Gelder *et al.* (1990) devised a strategy to linearly amplify mRNA from extremely small samples in their studies of gene expression in the brain. Their method, commonly referred to as the Eberwine method, has provided the basis of the procedures used today. The general steps involve reverse transcription of mRNA with an oligo dT primer, bearing a T7 RNA polymerase promoter site (*Figure 1*). After conversion of the mRNA-cDNA hybrid to double stranded complementary DNA (cDNA), antisense RNA (aRNA) is transcribed *in vitro* by T7 polymerase. The procedure can be repeated in a second round of amplification, further increasing the amplification factor, a necessity for samples in the lower nanogram range. Generally, the efficiency range for two rounds of linear amplification has been reported to lie between 10^3- and 10^5-fold amplification of the initial input amount. Note that cell line material and serial cell dilutions to obtain sequentially smaller samples yield better amplification efficiencies than clinical specimens. In the pre-microarray studies, this method was used even down to single cell analysis of gene expression of neurons in several studies (Cao *et al.*, 1996; Chow *et al.*, 1998; Crino *et al.*, 1996; Eberwine *et al.*, 1992; Phillips and Eberwine, 1996). In these studies, the complexity (the number of unique RNA sequences) of the aRNA products was used as a measure of success of amplification of a heterogeneous RNA pool.

4.1.1 Linear RNA amplification protocol

A protocol based on the original Eberwine and published by Baugh *et al.* (2001) is given in Protocol 1 at the end of this chapter. The final amplified RNA products are antisense. Care needs to be taken if oligo and not cDNA arrays are used in order to ensure that the correct strand is labelled. One option is to follow the protocol of an *in vitro* transcription kit that generates labelled aRNA ready for oligoarray hybridization in the second round of amplification.

4.1.2 Alternative global mRNA amplification strategies

The Eberwine-based RNA amplification method is 3′ end biased and complete coverage of the 5′ end is not ensured. For that particular reason, exploitation of a template switching effect inherent to the reverse transcriptase has been applied in an alternative protocol aimed at synthesizing full length cDNA (Wang *et al.*, 2000). This template switch effect is based on the terminal transferase activity of the reverse transcriptase enzyme that adds additional, non-template residues, primarily cytosines, to the 3′ end of the cDNA (*Figure 2*). The reaction mixture includes a primer containing an oligo dG sequence at its 3′ end which will base pair with the newly synthesized dCTP stretch. The reverse transcriptase then switches template and continues replicating the defined sequence of the annealed primer. The result is full length cDNA. The impact of 5′ end loss on the mRNA transcript is, however, dependent on the probe design which traditionally has been 3′ biased. Linear amplification is not restricted to the use of RNA polymerase. DNA polymerase is applied in a single primer isothermal amplification (SPIA) method adapted for microarray experiments (Smith *et al.*, 2003).

Hybridization to oligonucleotide arrays is strand specific and requires the antisense strand of nucleic acids. Two principal strategies have tackled this issue. One is to alter the classical protocol so that sense RNA is synthesized instead, and it can be

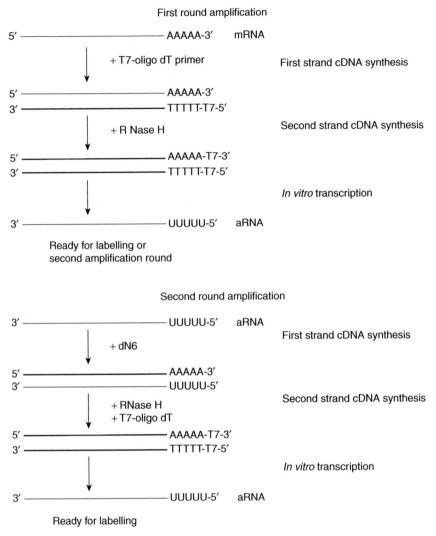

Figure 1. *Flowchart of a global, linear mRNA amplification procedure generating antisense RNA (aRNA). An oligo dT primer containing a T7 polymerase binding site is used to prime the first strand cDNA synthesis. Digestion of the mRNA strand in the mRNA-cDNA hybrid by RNase H leaves small fragments of RNA, which are used to prime second strand cDNA synthesis. Antisense RNA is then transcribed by T7 RNA polymerase. Second and subsequent rounds of amplification are initiated by random priming.*

noted that a number of protocol variations on this theme have been presented. The other strategy is to label and hybridize with aRNA. This latter alternative is the main approach, and standardized step regardless of sample size, in the user protocol connected with the widely used Affymetrix arrays.

The PCR-based exponential strategies were introduced to further increase the amplification factor, achieving factors as high as 3×10^{11} (Iscove *et al.*, 2002). The PCR-based

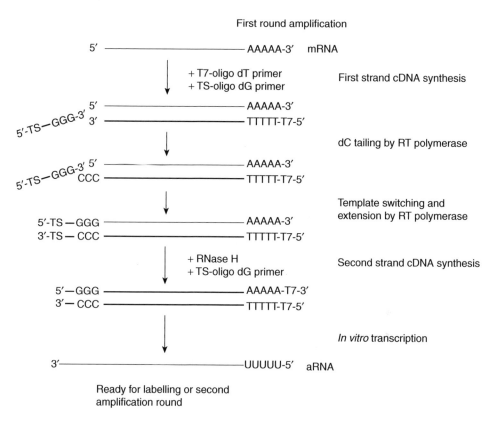

Figure 2. *Overview of a linear mRNA amplification based on the procedure described by Wang et al. (2000). Following oligo dT priming, the method exploits the template switching effect of the reverse transcriptase enzyme. The RT enzyme incorporates non-template dCTPs at the 3' end of the transcript, then switches templates and continues replication to the end of the primer. The result is full length cDNA. For the second strand, a primer with bases complementary to the dCTP stretch is applied. Antisense RNA is transcribed by the T7 RNA polymerase.*

methods share the feature of introducing PCR-priming sites at both ends of each reverse transcribed cDNA molecule, followed by global amplification of cDNA by PCR cycles (*Figure 3*). Again, we find protocols where the template-switching principle has been included, such as those using the SMART™ PCR technology. We can also find methods where the PCR step is followed by an *in vitro* transcription step, thereby combining the two principal strategies of mRNA amplification.

To follow the sample amplification protocols mentioned above, discrete reagents can be assembled by the investigator, or it is possible to choose between amplification kits from vendors such as Affymetrix, Arcturus, Ambion, Agilent Technologies, SuperArray Bioscience, Clontech, Telechem Int., Roche Applied Biosciences and NuGen (*Table 1*). *Table 1* displays information regarding the minimum input material amount specified in the respective manuals found on the manufacturers' web sites. In general, two rounds of amplification are required to generate sufficient amplified

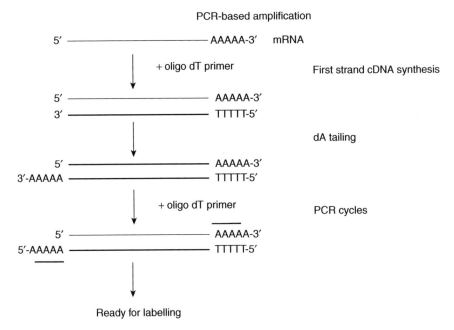

Figure 3. *Schematic illustration based on a reported exponential amplification method (Iscove et al., 2002). The mRNA transcript is reverse transcribed with an oligo dT primer. After RNA strand digestion the cDNA was tailed by terminal transferase to create an oligo dA tail. Addition of a poly-dA stretch allows the use of one oligo dT- (or oligo dT adaptor-) primer to be used in subsequent PCR cycles.*

RNA when starting with minimum total RNA quantities. A few manufacturers provide a minimum value of input material that is different from their recommended input, indicating that they have a sensitive assay, but the rate of success is variable for the lowest input range.

4.1.3 General consequences of mRNA amplification

As mentioned earlier, the conventional Eberwine-based amplification protocols involve directional priming from the 3′ ends of mRNA transcripts in the first round of amplification and random priming in the second and subsequent rounds, leading to shortened products. The final aRNA transcript lengths are dependent on the processive features of the specific reverse transcriptases and polymerases applied, as well as the reaction conditions. Product lengths after two rounds are typically in the range of 200–1000 bp, peaking at 400–500 bp. The 3′ end bias and lack of full-length products generally have a minimal effect in most microarray experiments, as probe sets (both cDNA and oligonucleotide probes) themselves are generally 3′ biased, as are conventional target-labelling protocols.

The aspect of reproducibility has been investigated for a number of global mRNA amplification procedures and is in general reported to be high. Parallel amplifications yield highly correlated expression profiles (Baugh *et al.*, 2001; Wang *et al.*, 2000).

Table 1. An overview of commercially available global RNA amplification kits.

Kit	Manufacturer	Principle	Range of input total RNA	Min. input total RNA	Recommended min. input total RNA
GeneChip Two Cycle Target Labeling	Manufactured by Invitrogen for Affymetrix	Linear amp	10–100 ng	10 ng	10 ng
RiboAmp	Arcturus	Linear amp	1– 40 ng	1–10 ng (250–500 cells)	10–40 ng (500–2000 cells)
RiboAmp HS	Arcturus	Linear amp	100 pg–1 ng	100–500 pg (10–50 cells)	500 pg–1 ng
MessageAmp II	Ambion	Linear amp	0.1–100 ng	0.1 ng	100 ng
Low RNA Input Fluorescent Linear Amplification	Agilent Technologies	Linear amp	50 ng–5 μg	50 ng	50 ng
TrueLabeling-PicoAMP	SuperArray Bioscience	Linear amp	50–500 pg	50 pg	50 pg
BD SMART mRNA Amplification	Clontech	Linear amp (Template switch mechanism)	0.1–5 μg	100 ng	100 ng
BD Atlas SMART Fluorescent Probe Amplification	Clontech	PCR-based (Template switch mechanism applied)	10 ng–1 μg	10 ng	10 ng
ArrayIt MiniAmp	TeleChem Inc	PCR-based (5–10 cycles) and one round linear amp	50 ng–1 μg	50 ng	50 ng
RAS Microarray Target Amplification	Roche	PCR-based	50 ng–1 μg	50 ng (1000 cells)	50 ng (1000 cells)
Ovation (Ribo-SPIA) Aminoallyl RNA Amplification	NuGen	Linear isothermal amp	5–100 ng	5 ng	5 ng

Abbreviations: amp (amplification).

In a comparison between linear and PCR-based amplification, the results showed that reproducibility was very high for linear amplification, and slightly lower for the SMART PCR-based amplification (Puskas *et al.*, 2002).

However, the most important aspect for any amplification protocol to be used in combination with quantitative analysis of gene expression is that the relative transcript abundance present in the initial mRNA sample is maintained throughout the procedure. The presence of transcript abundance bias during the scaling-up procedure will render the output data quantitatively unreliable. Systematic assessments of potential distortions to relative transcript abundance or other limitations of global mRNA amplification are therefore important steps before presenting data generated from amplified samples and drawing biological conclusions.

Despite the lack of such systematic assessments, sample amplification became a method of choice for profiling small samples. The Affymetrix platform (high density oligonucleotide arrays) integrated at an early stage the use of linear mRNA amplification as a standard step. In early reports introducing the global mRNA amplification for miroarray studies, the documentation of the degree of fidelity was generally conducted by comparing profiles between amplified and non-amplified material, comparing absent or present calls (Affymetrix platform), use of internal RNA standards, or northern blot or real-time reverse transcriptase-PCR (RT-PCR) verification. These evaluations covered linear-based procedures (Baugh *et al.*, 2001; Hu *et al.*, 2002; Scheidl *et al.*, 2002; Scherer *et al.*, 2003; Wang *et al.*, 2000; Zhao *et al.*, 2002), exponential-based procedures (Petalidis *et al.*, 2003; Seth *et al.*, 2003), or compared both of these strategies (Puskas *et al.*, 2002; Wang *et al.*, 2003). Evaluation of consistency of outliers between amplified and non-amplified material was a widely used parameter. The first two papers to quantitatively inspect the differences in gene expression measurements before and after linear amplification both concluded that the concordance was high, although there were some discrepancies that increased as the input of mRNA into the reaction decreased (Baugh *et al.*, 2001; Wang *et al.*, 2000). These observations have been confirmed in a number of published studies (Feldman *et al.*, 2002; Hu *et al.*, 2002; Puskas *et al.*, 2002; Scheidl *et al.*, 2002; Zhao *et al.*, 2002). It is worth mentioning that in the studies of amplification fidelity, the authors generally confer a subjective opinion of what defines sufficient amplification accuracy, and in several cases the analysis is limited to a subgroup of genes. Nonetheless, the general conclusion drawn from these studies was that microarray data from amplified material is comparable to non-amplified material, but that there is a slight decrease in correlation coefficients, reflecting changes in transcript ratios. For moderate input total RNA amounts (above nanogram range), microarray analysis of amplified targets has been shown to have a number of advantages compared with the use of conventional non-amplified targets, and has been applied in a number of published studies, even where samples were sufficient for non-amplified target preparation.

4.1.4 Lower limit considerations

Technically, it has been demonstrated that the efficiency of certain amplification protocols for mRNA even allow single-cell profiling. The typical total RNA yield from one mammalian cell is 10–30 pg. Hence: Is the use of consecutive linear amplification rounds or PCR cycles the answer to profile minute samples? Many studies have found a markedly reduced correlation with extremely small samples, especially for mRNA transcripts in the low abundance range (Baugh *et al.*, 2001; Scheidl *et al.*, 2002; Wang *et al.*, 2000). A common feature for studies employing minute samples is the reduced number of gene-specific transcripts detected on the arrays (Mohr *et al.*, 2004; Nakagawa and Schwartz, 2004; Nygaard *et al.*, 2005). Few investigators have established the lower boundaries with respect to amplification fidelity, but report that variability is increased in experiments with low RNA input values. In fact, questionable reliability of quantitative measures from low abundant mRNA transcripts is an issue that affects all of the gene expression technologies mentioned, when starting with a highly diluted and complex mRNA template mix. Evidence published by Nygaard *et al.* (2005) demonstrated that in a microarray setting, the quantitative accuracy of expression measurements was greatly affected by stochastic effects exerted on low abundant mRNA molecules. The lower the abundance of any template, the smaller the probability that its true abundance will be maintained in the amplified product (Stenman *et al.*, 2003). This implies that the yield of quantitative data from scarce material is restricted to a few highly expressed genes. The analysis approach by Nygaard *et al.* (2005) allowed the investigators to pinpoint the gene-specific number of transcript templates that defined the limit of reliability with respect to the number of cells from that particular source. From the experiment amplifying from 1000 cells, transcripts expressed with at least 121 transcripts per cell were statistically reliable, and for 250 cells, the limit was 1806 transcripts per cell. In the 250-cell sample size, fewer than 400 genes were left that were reliable for further biological interpretation. These 400 genes represented about 5% of the average number of genes registered on the particular replicate arrays. Hence, the variance or noise of gene expression measurements is substantial for minute samples and clearly questions the reliability of establishing a transcriptome overview of any single cells or even tens of cells. Because of the large portion of low-expressing genes, a large number of measurements are hampered by large variability, leading to difficulties in finding, tracking or deciphering gene expression patterns/signatures involving subsets of genes. Partial qualitative information can, however, be retained, such as the presence of certain transcripts, typically high copy number transcripts. Validation of microarray data is generally performed using qRT-PCR. However, as mentioned above, this technique suffers from the same inconsistencies at low copy numbers and may similarly not represent true measurements. These features require the investigator to fully understand the risk and take the consequences by filtering out unreliable data so as not to confer biological significance on invalid quantitative data. Taking these aspects into account implies the restriction of microarray experiments based on amplified material from certain laser microdissected cell samples, single-cell picking and other samples with low numbers of cells.

If we return to the theme of the transcriptome of a sample, however, there is an additional aspect of stochastic perturbation to have in mind when looking into the transcriptome of single cells. In addition to stochastic effects exerted on the templates during global sample amplification, investigators are also confronted by stochastic aspects of gene expression. Stochastic fluctuations are considered to be significant in small systems, where components are present at very low concentrations (Swain *et al.*, 2002). These processes introduce random variation in gene expression among supposedly identical cells, which is referred to as gene expression noise (Raser and O'Shea, 2005). The detection of illegitimate transcripts (low-level transcription of tissue-specific gene in non-specific cells, see Chelly *et al.*, 1989) exemplifies the presence of genes noise.

Hence, the extraction of biological significance of gene expression data obtained from amplified single cell material is burdened by stochastic effects both on cellular gene expression and on the global transcript amplification process. The difficulties lie in distinguishing genuine, reliable transcription levels from stochastically generated noise. Furthermore, we do not really know how much heterogeneity to expect between the transcriptome profile of one cell compared with another, further complicating the matter.

With these complicating issues regarding global mRNA amplification of tens of, or single, cell samples, perhaps we need to consider alternative approaches to reveal the gene activity in the cell/s of interest. DNA-based comparative genomic hybridization on arrays (array-cgh) has also faced the issue of excluding a number of potential sample types because of inadequate amounts of DNA to perform the analysis. Whole-genome amplification (WGA) methods have therefore been developed with the aim of generating sufficient DNA yields with minimal disturbance of the original profile. The three main strategies of WGA include: multiple displacement amplification (MDA; Dean *et al.*, 2002; Lage *et al.*, 2003), primer extension preamplification (PEP; Zhang *et al.*, 1992) and degenerate oligonucleotide primed PCR (DOP; Telenius *et al.*, 1992). Applications of these methods have included genomic DNA amplification from single cells for array-cgh analysis. Recent articles have assessed the degree of distortion linked to the amplification procedures (Arriola *et al.*, 2007; Paez *et al.*, 2004; Pinard *et al.*, 2006). Hence, as with global mRNA amplification, WGA is associated with experimental variability affecting the output data. We know from previous experience working with global mRNA amplification that biological and experimental variability are more tightly entwined when initiating amplification from minute RNA amounts. However, it is worth recalling that a major difference between the comparative analysis of the transcriptome and the genome is the variable or constant signal baseline of the reference sample. The reference or control value in the RNA-based analysis can lie anywhere from zero transcript copies per cell to 10 000 per cell, while for DNA analysis the reference comprises the normal state of a diploid cell, in other words two copies per cell. Furthermore, in RNA-based analysis, probes are designed to detect one specific target in order to quantify the expression level of a particular gene. In contrast, for genomic analysis, each DNA region we quantify is covered by multiple probes along the region. For example, a DNA amplicon covering 30 Mb may be covered by 10 individual probes designed towards different targets but measuring the same genomic amplification. Signals from each of the probes contribute with an independent and equal quantitative count. Hence a stochastic component affecting any one of the measurements may be easier to isolate, but still leave data to quantify the copy number level of the DNA region surveyed. For the reasons just mentioned, the biological and experimental variables are somewhat relaxed in DNA-based array analysis and with respect to performing analysis of only a few cells, could provide a more robust approach to assess levels of gene activity.

5   Concluding remarks

In summary, technological advances have allowed high-throughput analysis of gene expression at all levels of gene expression from DNA to RNA to protein. RNA-based microarray technology represents a widely applied technology and is the source of huge amounts of gene expression profiling data. There are a number of options regarding how to sample and prepare the RNA for microarray experiments which may influence the output data and therefore should be thoroughly considered.

Global mRNA amplification procedures have been developed to alleviate the large material requirements of microarray experiments. To date, a substantial number of biological investigations have included mRNA amplification prior to expression analysis and the popular Affymetrix oligoarray platform has incorporated linear mRNA amplification as a standard step in its experimental procedure. Sample amplification is often the only option for performing high-throughput microarray analysis on very small samples and should not necessarily be avoided, but the trade-off with time, assay cost and potentially the short list of reliable measurements, may be negative below a certain threshold of input material. For more moderate samples, RNA amplification represents a valid approach, with acceptable levels of transcript abundance preservation, and is making its way as a conventional procedure, even for large samples where amplification is not necessary.

Protocol 1: Linear RNA amplification

The protocol presented here is based on the original Eberwine and published by Baugh *et al.* (2001), further modified here.

Suggested commercial products and/or providers are in parenthesis.

Material and reagents
RNA *material*
Total RNA diluted to the appropriate working concentration

Reagents
Deoxynucleoside triphosphate (dNTP) mix solution
RNAsin (Promega)
Reverse transcriptase, buffer and dithiothreitol (DTT) (e.g. Superscript from Invitrogen or Powerscript from Clontech)
T4gp32 protein (USB)
DNA polymerase and buffer (e.g. Taq polymerase from Promega)
Rnase H (Gibco BRL, Life Technologies)
E.coli ligase (New England Biolabs)
$MgCl_2$
T4 polymerase (New England Biolabs)
Chloroform
Phenol/chloroform (1:1)
96% and 70% ethanol
7.5 M NH_4Ac
Linear polyacrylamide (Ambion)
In vitro transcription kit (e.g. Megascript from Ambion)
Trizol (Gibco BRL, Life Technologies)
Isopropanol

Primers
Oligo dT-T7 primer (5'-GCA-TTA-GCG-GCC-GCG-AAA-TTA-ATA-CGA-CTC-ACT-ATA-GGG-AGA-$T_{(21)}$-V-3') (Baugh *et al.*, 2001)
Random hexamer primers (dN6) (Roche Diagnostics)

Tubes and columns
 0.2 ml and 1 ml eppendorf tubes
 2 ml phase lock tubes
 Micro Bio spin columns, P6, (Bio-Rad Laboratories)

Apparatus
 PCR apparatus
 Speed vacuum

Procedure

First round of amplification
First strand cDNA synthesis

1a. In a 0.2 ml PCR reaction tube, mix 200 ng total RNA with 100 ng oligo dT/T7 primer in a total volume of 5 µl. Heat to 70°C for 4 min, then snap cool on ice.

1b. Add 5 µl first strand cDNA synthesis mix (2 µl 5 × 1st strand buffer, 1 µl 0.1 M DTT, 0.5 µl 10 mM dNTP, 0.5 µl RNasin (~20 U), 0.5µl (8 mg ml^{-1}) T4gp32, 0.5µl RT enzyme (100 U)). Incubate at 42°C for 1 h in a thermal cycler, then heat inactivate at 65°C for 15 min. Chill on ice.

Second strand cDNA synthesis

2a. Prepare a 2nd strand synthesis mix with a volume of 65 µl (15 µl 5 × 2nd strand buffer, 1.5 µl 10 mM dNTP, 20 U DNA polymerase I, 1 U RNase H, 5 U *E.coli* ligase, and dH$_2$O to a final 65 µl final volume). Incubate at 16°C for 2 h in a thermal cycler.

2b. Add 10 U T4 DNA polymerase and incubate at 16°C for 15 min.

2c. Heat inactivate at 70°C for 10 min.

cDNA cleanup

3a. Transfer cDNA mix to 1 ml tube or phase lock tube. Add 75 µl phenol/chloroform, mix well then spin for 5 min at 12 000 g.

3b. Transfer aqueous phase to a prepared bio-spin 6 column.

3c. Add 1 µl linear acrylamide, 35 µl 7.5 M NH$_4$Ac and 500 µl 96% ethanol to flow through and precipitate cDNA for at least 2 hours at −20°C.

3d. Collect cDNA pellet by centrifugation (30 min) and carefully remove supernatant. Wash twice with 96% ethanol, air dry and re-dissolve in 9 µl dH$_2$O.

In vitro *transcription*

4. Prepare *in vitro* transcription mix as suggested by manufacturer (e.g. Megscript, Ambion: 2 µl of each 75 mM dNTP, 2 µl reaction buffer, 2 µl enzyme mix) and add 8 µl cDNA for a total volume of 20 µl in a 1 ml tube. Incubate for 3 h at 37°C.

aRNA cleanup

5a. Any manufactured RNA isolation kit can be applied or Trizol as we suggest here.

5b. Transfer *in vitro* mix to phase lock tube. Add 1 ml Trizol reagent, mix well then add 0.2 ml chloroform and mix well again. Allow tube to stand for 2–3 min at room temperature.

5c. Centrifuge for 15 min at 4°C and transfer aqueous phase to a fresh tube. Add 0.5 ml isopropanol and 1 µl linear polyacrylamide. Let the aRNA precipitate overnight at –20°C.

5d. Centrifuge at 12 000 g for 30 min and remove discard supernatant. Wash twice with 70% ethanol, carefully air-dry and re-dissolve in 10 µl dH$_2$O.

Second round of amplification

First strand cDNA synthesis

6a. Add 0.5 µl (1 µg/µl^{-1}) dN6 to aRNA from first round and reduce total volume to 5 µl using a speed vacuum. Transfer to a 0.2 ml PCR tube and heat to 70°C for 4 min, then snap cool on ice.

6b. Make same first strand cDNA synthesis mix as in first round and incubate at the following temperatures in a thermal cycler; 37°C for 20 min, 42°C for 20 min, 50°C for 10 min, 55°C for 10 min, 65°C for 15 min.

6c. Add 1 U of RNase H and incubate at 37°C for 30 min, followed by 95°C for 2 min and snap cool on ice and give a short spin.

Second strand cDNA synthesis

7a. For second strand priming, add 100 ng oligo dT/T7 primer and anneal at 42°C for 10 min and chill again on ice.

7b. Prepare and add the same second strand synthesis mix as in the first round of amplification (minus ligase). Incubate at 16°C for 2 h in a thermal cycler.

7c. Add 10 U T4 DNA polymerase and incubate at 16°C for 15 min.

7d. Heat inactivate at 70°C for 10 min.

cDNA cleanup

8. Purify double-stranded cDNA as in the first round.

In vitro *transcription and aRNA cleanup*

9a. Follow same procedure as in first round of amplification but dissolve final aRNA in 20 µl.

9b. Quantify aRNA and quality control aRNA.

References

Arriola, E., Lambros, M.B., Jones, C. *et al.* (2007) Evaluation of Phi29-based whole-genome amplification for microarray-based comparative genomic hybridisation. *Lab. Invest.* **87**: 75–83.

Baugh, L.R., Hill, A.A., Brown, E.L. & Hunter, C.P. (2001) Quantitative analysis of mRNA amplification by in vitro transcription. *Nucleic Acids Res.* **29**: E29.

Bertone, P. & Snyder, M. (2005) Advances in functional protein microarray technology. *Febs. J.* **272**: 5400–5411.

Bishop, J.O., Morton, J.G., Rosbash, M. & Richardson, M. (1974) Three abundance classes in HeLa cell messenger RNA. *Nature* **250**: 199–204.

Cao, Y., Wilcox, K.S., Martin, C.E., Rachinsky, T.L., Eberwine, J. & Dichter, M. A. (1996) Presence of mRNA for glutamic acid decarboxylase in both excitatory and inhibitory neurons. *Proc. Natl Acad. Sci. USA* **93**: 9844–9849.

Chelly, J., Concordet, J.P., Kaplan, J.C. & Kahn, A. (1989) Illegitimate transcription: transcription of any gene in any cell type. *Proc. Natl Acad. Sci. USA* **86**: 2617–2621.

Chow, N., Cox, C., Callahan, L.M., Weimer, J.M., Guo, L. & Coleman, P.D. (1998) Expression profiles of multiple genes in single neurons of Alzheimer's disease. *Proc. Natl Acad. Sci. USA* **95**: 9620–9625.

Crino, P.B., Trojanowski, J.Q., Dichter, M.A. & Eberwine, J. (1996) Embryonic neuronal markers in tuberous sclerosis: single-cell molecular pathology. *Proc. Natl Acad. Sci. USA* **93**: 14152–14157.

Darzacq, X., Singer, R.H. & Shav-Tal, Y. (2005) Dynamics of transcription and mRNA export. *Curr. Opin. Cell Biol.* **17**: 332–339.

Dean, F.B., Hosono, S., Fang, L. *et al.* (2002) Comprehensive human genome amplification using multiple displacement amplification. *Proc. Natl Acad. Sci. USA* **99**: 5261–5266.

Dobbin, K., Shih, J.H. & Simon, R. (2003) Questions and answers on design of dual-label microarrays for identifying differentially expressed genes. *J. Natl Cancer Inst.* **95**: 1362–1369.

Eberwine, J., Yeh, H., Miyashiro, K., Cao, Y., Nair, S., Finnell, R., Zettel, M. & Coleman, P. (1992) Analysis of gene expression in single live neurons. *Proc. Natl Acad. Sci. USA* **89**: 3010–3014.

Feldman, A.L., Costouros, N.G., Wang, E., Qian, M., Marincola, F.M., Alexander, H.R. & Libutti, S.K. (2002) Advantages of mRNA amplification for microarray analysis. *Biotechniques* **33**: 906–912, 914.

Hastie, N.D. & Bishop, J.O. (1976) The expression of three abundance classes of messenger RNA in mouse tissues. *Cell* **9**: 761–774.

Hu, L., Wang, J., Baggerly, K., Wang, H., Fuller, G.N., Hamilton, S.R., Coombes, K.R. & Zhang, W. (2002) Obtaining reliable information from minute amounts of RNA using cDNA microarrays. *BMC Genomics* **3**: 16.

Iscove, N.N., Barbara, M., Gu, M., Gibson, M., Modi, C. & Winegarden, N. (2002) Representation is faithfully preserved in global cDNA amplified exponentially from sub-picogram quantities of mRNA. *Nat. Biotechnol.* **20**: 940–943.

Karsten, S.L., Van Deerlin, V.M., Sabatti, C., Gill, L.H. & Geschwind, D.H. (2002) An evaluation of tyramide signal amplification and archived fixed and frozen tissue in microarray gene expression analysis. *Nucleic Acids Res.* **30**: E4.

Kendziorski, C., Irizarry, R.A., Chen, K.S., Haag, J.D. & Gould, M.N. (2005) On the utility of pooling biological samples in microarray experiments. *Proc. Natl Acad. Sci. USA* **102**: 4252–4257.

Lage, J.M., Leamon, J.H., Pejovic, T. *et al.* (2003) Whole genome analysis of genetic alterations in small DNA samples using hyperbranched strand displacement amplification and array-CGH. *Genome Res.* **13**: 294–307.

Lockwood, W.W., Chari, R., Chi, B. & Lam, W.L. (2006) Recent advances in array comparative genomic hybridization technologies and their applications in human genetics. *Eur. J. Hum. Genet.* **14**: 139–148.

Maercker, C. (2005) Protein arrays in functional genome research. *Biosci. Rep.* **25**: 57–70.

Mattick, J.S. (2001) Non-coding RNAs: the architects of eukaryotic complexity. *EMBO Rep.* **2**: 986–991.

Mohr, S., Bottin, M.C., Lannes, B., Neuville, A., Bellocq, J.P., Keith, G. & Rihn, B.H. (2004) Microdissection, mRNA amplification and microarray: a study of pleural mesothelial and malignant mesothelioma cells. *Biochimie* **86**: 13–19.

Nakagawa, T. & Schwartz, J.P. (2004) Gene expression profiles of reactive astrocytes in dopamine-depleted striatum. *Brain Pathol.* **14**: 275–280.

Nygaard, V., Holden, M., Loland, A., Langaas, M., Myklebost, O. & Hovig, E. (2005) Limitations of mRNA amplification from small-size call samples. *BMC Genomics* **6**: 147.

Nygaard, V., Liu, F., Holden, M., Kuo, W.P., Frigessi, A., Glad, I., van de Wiel, M., Hovig, E. & Lyng, H. (manuscript submitted) Consistency of absolute transcript concentrations across high throughput technologies.

Paez, J. G., Lin, M., Beroukhim, R. *et al.* (2004) Genome coverage and sequence fidelity of phi29 polymerase-based multiple strand displacement whole genome amplification. *Nucleic Acids Res.* **32**: e71.

Petalidis, L., Bhattacharyya, S., Morris, G.A., Collins, V.P., Freeman, T.C. & Lyons, P.A. (2003) Global amplification of mRNA by template-switching PCR: linearity and application to microarray analysis. *Nucleic Acids Res.* **31**: e142.

Phillips, J. & Eberwine, J.H. (1996) Antisense RNA amplification: A linear amplification method for analyzing the mRNA population from single living cells. *Methods* **10**: 283–288.

Pinard, R., de Winter, A., Sarkis, G.J., Gerstein, M.B., Tartaro, K.R., Plant, R. N., Egholm, M., Rothberg, J.M. & Leamon, J.H. (2006) Assessment of whole genome amplification-induced bias through high-throughput, massively parallel whole genome sequencing. *BMC Genomics* **7**: 216.

Puskas, L.G., Zvara, A., Hackler, L., Jr. & Van Hummelen, P. (2002) RNA amplification results in reproducible microarray data with slight ratio bias. *Biotechniques* **32**: 1330–1334, 1336, 1338, 1340.

Quinlan, T.J., Beeler, G.W., Jr., Cox, R.F., Elder, P.K., Moses, H.L. & Getz, M.J. (1978) The concept of mRNA abundance classes: a critical re-evaluation. *Nucleic Acids Res.* **5**: 1611–1625.

Raser, J.M. & O'Shea, E.K. (2005) Noise in gene expression: origins, consequences, and control. *Science* **309**: 2010–2013.

Reverter, A., McWilliam, S.M., Barris, W. & Dalrymple, B.P. (2005) A rapid method for computationally inferring transcriptome coverage and microarray sensitivity. *Bioinformatics* **21**: 80–89.

Scheidl, S.J., Nilsson, S., Kalen, M., Hellstrom, M., Takemoto, M., Hakansson, J. & Lindahl, P. (2002) mRNA expression profiling of laser microbeam microdissected cells from slender embryonic structures. *Am. J. Pathol.* **160**: 801–813.

Schena, M., Shalon, D., Davis, R.W. & Brown, P.O. (1995) Quantitative monitoring of gene expression patterns with a complementary DNA microarray. *Science* **270**: 467–470.

Scherer, A., Krause, A., Walker, J.R., Sutton, S.E., Seron, D., Raulf, F. & Cooke, M.P. (2003) Optimized protocol for linear RNA amplification and application to gene expression profiling of human renal biopsies. *Biotechniques* **34**: 546–550, 552–554, 556.

Seth, D., Gorrell, M.D., McGuinness, P.H., Leo, M.A., Lieber, C.S., McCaughan, G.W. & Haber, P.S. (2003) SMART amplification maintains representation of relative gene expression: quantitative validation by real time PCR and application to studies of alcoholic liver disease in primates. *J. Biochem. Biophys. Methods* **55**: 53–66.

Smith, L., Underhill, P., Pritchard, C. *et al.* (2003) Single primer amplification (SPA) of cDNA for microarray expression analysis. *Nucleic Acids Res.* **31**: e9.

Stears, R.L., Getts, R.C. & Gullans, S.R. (2000) A novel, sensitive detection system for high-density microarrays using dendrimer technology. *Physiol. Genomics* **3**: 93–99.

Stenman, J., Lintula, S., Rissanen, O., Finne, P., Hedstrom, J., Palotie, A. & Orpana, A. (2003) Quantitative detection of low-copy-number mRNAs differing at single nucleotide positions. *Biotechniques* **34**: 172–177.

Swain, P.S., Elowitz, M.B. & Siggia, E.D. (2002) Intrinsic and extrinsic contributions to stochasticity in gene expression. *Proc. Natl Acad. Sci. USA* **99**: 12795–12800.

Szaniszlo, P., Wang, N., Sinha, M., Reece, L.M., Van Hook, J.W., Luxon, B.A. & Leary, J.F. (2004) Getting the right cells to the array: Gene expression microarray analysis of cell mixtures and sorted cells. *Cytometry A* **59**: 191–202.

Telenius, H., Carter, N.P., Bebb, C.E., Nordenskjold, M., Ponder, B.A. & Tunnacliffe, A. (1992) Degenerate oligonucleotide-primed PCR: general amplification of target DNA by a single degenerate primer. *Genomics* **13**: 718–725.

Ueda, H.R., Hayashi, S., Matsuyama, S., Yomo, T., Hashimoto, S., Kay, S.A., Hogenesch, J.B. & Iino, M. (2004) Universality and flexibility in gene expression from bacteria to human. *Proc. Natl Acad. Sci. USA* **101**: 3765–3769.

Van Gelder, R.N., von Zastrow, M.E., Yool, A., Dement, W.C., Barchas, J.D. & Eberwine, J.H. (1990) Amplified RNA synthesized from limited quantities of heterogeneous cDNA. *Proc. Natl Acad. Sci. USA* **87**: 1663–1667.

Velculescu, V. E., Madden, S. L., Zhang, L. *et al.* (1999) Analysis of human transcriptomes. *Nat. Genet.* **23**: 387–388.

Wang, E., Miller, L.D., Ohnmacht, G.A., Liu, E.T. & Marincola, F.M. (2000) High-fidelity mRNA amplification for gene profiling. *Nat. Biotechnol.* **18**: 457–459.

Wang, J., Hu, L., Hamilton, S.R., Coombes, K.R. & Zhang, W. (2003) RNA amplification strategies for cDNA microarray experiments. *Biotechniques* **34**: 394–400.

Zhang, L., Cui, X., Schmitt, K., Hubert, R., Navidi, W. & Arnheim, N. (1992) Whole genome amplification from a single cell: implications for genetic analysis. *Proc. Natl Acad. Sci. USA* **89**: 5847–5851.

Zhang, S.D. & Gant, T.W. (2005) Effect of pooling samples on the efficiency of comparative studies using microarrays. *Bioinformatics* **21**: 4378–4383.

Zhao, H., Hastie, T., Whitfield, M.L., Borresen-Dale, A.L. & Jeffrey, S.S. (2002) Optimization and evaluation of T7 based RNA linear amplification protocols for cDNA microarray analysis. *BMC Genomics* **3**: 31.

3

Methods of proteome analysis: Challenges and opportunities

Sarah R. Hart and Simon J. Gaskell

1 Proteomics and the quantitative challenge

An integrative approach to the investigation of the multicellular systems in biology requires the study of multiple 'omic' datasets, to develop an integrated model of cellular behaviour using disparate genomic, transcriptomic, metabolomic and proteomic data. Of these, the proteome is arguably the most dynamic, highly regulated and complex '-ome'. Regulated protein expression, alternate splicing and internal start sites, modification and degradation, all diversify the nature of individual proteins within proteomes. All this detail must be understood if we are to model the cell accurately, fully understand its behaviour, and how disruptions to normal behaviour leads to disease. Proteins act as catalysts for many biological processes, and their concentration and catalytic activity accordingly determine the rate of biochemical reactions. The precise and accurate determination of the concentration of specific protein components will become increasingly important in systems approaches to biology. Whilst this chapter cannot be entirely comprehensive in scope, it introduces the pertinent issues in qualitative and quantitative proteomics within the context of systems biology, presenting both striking progress and areas where significant challenges remain.

Proteomics is rapidly becoming a mature approach, yet many technical challenges remain. The most important of these arises from the inherent complexity of the proteome. Numerous different proteins must be characterized to fully determine the proteome, yet the sheer complexity of the proteome does not arise simply from the range of different genes expressed within a given cell system. Additional complexity arises from a large number of sources, which can be pre-, co-, or post-translational in origin. In order for the cell to perform its correct functions, the expression, behaviour and degradation of proteins are highly regulated, with numerous mechanisms contributing to this tight regulation.

In the first place, alternate splicing of exons from pre-mRNA to give different transcripts results in multiple translated proteins generated from the DNA code. The mechanisms regulating alternate splicing remain to be fully characterized, but are thought to contribute significantly to diseases such as cancer (Blencowe, 2006; Skotheim and Nees, 2007), and alternate splicing is believed to contribute significantly to the functional complexity observed in the human proteome (Brett *et al.*, 2002; Lee and Roy, 2004). Additional complexity arises from the presence of translational start sites in sub-optimal contexts, which can lead to the generation of alternate translational start sites. In many cases, this allows the expressed proteins thus generated to take up different subcellular locations, having different lifetimes and different functional characteristics (Kochetov 2006; Oyama *et al*., 2007). By far the most abundant and highly-regulated mechanism for altering the functional behaviour of proteins is post-translational modification (PTM). Current estimates indicate that the number of modifications within the proteome as a whole may approach as much as one modification per amino acid (Nielsen *et al*., 2006; see Garavelli 2004 for a comprehensive list of modifications). The most abundant amongst the post-translational modifications is *O*-phosphorylation, with approximately 30% of proteins containing covalently-bound phosphate (Cohen, 2000). The primary sites of phosphorylation are serine, threonine and tyrosine residues, with estimated relative abundances of 1800:200:1 (Hunter, 1998; Olsen *et al*., 2006). Targeted proteomic methods for the isolation and discovery of phosphoproteins are discussed later in this chapter. A wide variety of other PTMs (including glycosylation, acetylation, ubiquitination and many others) further increase the observed complexity, and have a significant effect upon the sensitivity and dynamic range that are achievable with proteome analyses.

The diversity of the proteome therefore represents an enormous challenge to both qualitative and quantitative analyses. Comprehensive determination of a proteome – encompassing all the qualitative and quantitative aspects that are discussed in the sections below – remains an elusive goal, though one that is increasingly approached using the experimental methods that are discussed in the remainder of this chapter.

2 Qualitative proteome analyses

2.1 *Common analytical workflows*

The vast majority of proteomic studies are entirely qualitative in nature; that is, they seek to characterize the nature of the components present in biological samples. All modern proteomics methods rely on the application of mass spectrometry (MS), most commonly to analyse peptides derived from the proteins of interest by enzyme-mediated hydrolysis, rather than the intact proteins themselves. The analytical strategies adopted can broadly be classified in terms of the primary methods for separation of the analytes. The first and more traditional set of methods involves separation of proteins upon the basis of their physicochemical properties, most commonly by electrophoretic methods. The 1-D variant of the method achieves crude separation according to molecular weight, whereas 2-D gel electrophoresis additionally separates according to protein isoelectric point. Alternatively, the entire proteome can be subjected to site-specific proteolysis and the proteolytic peptides thus generated separated for analysis; this is the so-called 'shotgun proteomics' approach. Each approach has specific benefits and some disadvantages.

Gel electrophoresis-based approaches have the advantage that all peptides derived from a given protein thus separated should be contained within a specific gel band or spot (Carrette *et al*., 2006), lending themselves well to conventional MS analyses and permitting facile bioinformatic data interpretation (as discussed below). Proteins within polyacrylamide gels can be visualized on the basis of their staining with silver (Blum and Gross, 1987; Gharahdaghi *et al*., 1999; Shevchenko *et al*., 1996), Coomassie (Schagger *et al*., 1988) or with fluorescent staining methods (Lopez *et al*., 2000), some of which can enable relative quantification (Unlu *et al*., 1997). Problems arise from the loading capacity of 2-D gels, commercial implementations of which limit capacity to approximately 1–3 mg in the first dimension; this in turn limits the dynamic range which is achievable in such an analysis. The resolution achievable on a standard format 2-D gel is high; in principle many hundreds of components can be separated, including in some cases post-translationally modified variants of the same protein. It is not unusual, however, to find two or more components per gel spot, confounding quantification on the basis of spot stain intensity. In addition, proteins with extreme properties, for example very large or small molecular weight, extremes of pI and highly hydrophobic proteins, are poorly represented on standard 2-D gels (Lopez, 2007).

The main alternative to gel-based approaches is the use of multidimensional peptide chromatography, the so-called 'shotgun' approaches. The earliest applications of this method were developed by Yates and co-workers, who coined the term multidimensional protein identification technology (MudPIT; Washburn et al., 2001). These methods rely on denaturation of the protein mixture and site-specific proteolysis to generate peptides, which are then subjected to multiple (usually two) orthogonal stages of chromatographic separation. Typically the first dimension of separation uses strong cation exchange chromatography, separating peptides on the basis of charge, although other modes have been used. The second dimension is almost invariably separation on the basis of hydrophobicity by reversed-phase chromatography. Connectivity between the proteins and their proteolytic peptides is lost in this kind of approach, necessitating tandem mass spectrometric (MS/MS) analysis (as discussed below). Furthermore, identification of proteins requires that all data from an entire 2-D chromatographic run must be combined, as peptides from a single protein will be separated in both dimensions.

A third major strategy, the 'GeLC-MS' approach, is essentially a hybrid of the first two. Here, 1-D polyacrylamide gels are used to separate proteins, the entire gel lane is excised as small slices and reversed-phase liquid chromatographic separation and tandem mass spectrometric analyses are performed on the digested peptides from individual slices (Broadhead et al., 2006; Nicholas et al., 2006). This method has the advantage of providing a means to solubilize hydrophobic (such as membrane) proteins prior to separation using the gel loading buffer (Blonder, 2004). Since all peptides from a given protein remain within a single band, peptide connectivity with the progenitor proteins is maintained, albeit in the context of a possibly complex mixture analysis which necessarily limits the achievable dynamic range.

2.2 Protein recognition using conventional mass spectrometric data

The term protein *identification* is frequently used in proteome analyses; it is in fact somewhat misleading. What is generally meant is that evidence has been adduced for the expression of a particular gene. The reason for this distinction – which is analytically and biologically important – is that very commonly the mass spectrometric evidence obtained in proteome analyses represents only a proportion of each protein sequence. Accordingly, variations in structure (such as post-translational modifications) that may be present elsewhere in the sequence are not characterized. The term protein *recognition* represents a convenient shorthand to make this point and is used in the remainder of this chapter.

Recognition of proteins based upon conventional MS of hydrolysates is usually termed peptide mass fingerprinting (PMF; Henzel *et al.*, 1993, 2003; Pappin *et al.*, 1993), and is typically performed using matrix-assisted laser desorption/ionization (MALDI) time-of-flight (TOF) MS, which produces singly protonated peptides. The principle underlying PMF relies upon the use of specific endoproteinases, with predictable, sequence-specific cleavage properties, to generate peptides. Mass spectra from these peptides can then be matched to theoretical masses generated from sequence databases using algorithms (see *Figure 1*); some publicly available tools can perform this function (see *Table 1*), and in addition many instrument vendors have their own software implementations for PMF. The working assumption underpinning PMF is connectivity between ions, that is, that the peptide ions observed within a given mass spectrum originate from a common protein progenitor. This will often be the case with 1-D gel or 2-D gel separation of simple mixtures; however, as samples of increasing complexity are analysed, the effectiveness of PMF approaches declines, with mixtures containing more than 3–4 components being very difficult to define (Henzel *et al.*, 2003). When using peptide mass data alone, there are a number of ways in which the statistical confidence of protein recognitions can be improved. These include limited composition or sequence data derived from simple peptide derivatization procedures (Brancia *et al.*, 2001; Warwood *et al.*, 2006), and the use of high mass accuracy and highly reproducible chromatographic elution profiles (Strittmatter *et al.*, 2003), an approach which originally derived as an extension from LC-MS/MS approaches. Modest increases in the information yield of such an experiment thus generated can have significant dividends in the confidence of the resultant identifications, and are anticipated to be of particular utility where sample quantity is very limited.

2.3 *Protein recognition using tandem MS data*

A common means to increase the confidence of matching peptide mass spectrometric data to sequence databases is via the use of tandem mass spectrometry (MS/MS), which may be coupled with MALDI or with electrospray ionization (the latter commonly generating multiply protonated peptides). Gas-phase fragmentation of protonated peptide precursor ions by collision with inert gas molecules (collisionally activated dissociation, CAD) results in reproducible cleavage at peptide bonds (Biemann, 1988; Paizs and Suhai, 2005). The product ions thus generated are termed according to their origin; if derived from the C-terminal region of the peptide, the products are termed y ions, and if primary products derive from the N-terminus, they are termed b ions (Biemann, 1988; Roepstorff and Fohlman 1984; see *Figure 2*). Whilst in general CAD of protonated peptides generates series of b and y ions, some promotion of the cleavage of specific peptide bonds may be observed. Cleavage of N-terminal to proline residues is generally favoured, for example, (Paisz and Suhai, 2005; Schwartz and Bursey, 1992; Vaisar and Urban, 1996), whilst fragmentation of singly protonated peptides C-terminal to aspartic acid (and to a lesser extent, glutamic acid) residues is also commonly observed (Paisz and Suhai, 2005; Yu *et al.*, 1993).

The high degree of predictability (at least in a qualitative sense) of gas-phase peptide ion fragmentation has resulted in the rapid adoption of MS/MS as the predominant method for inference of peptide (and thus protein) identity within proteomics, and to

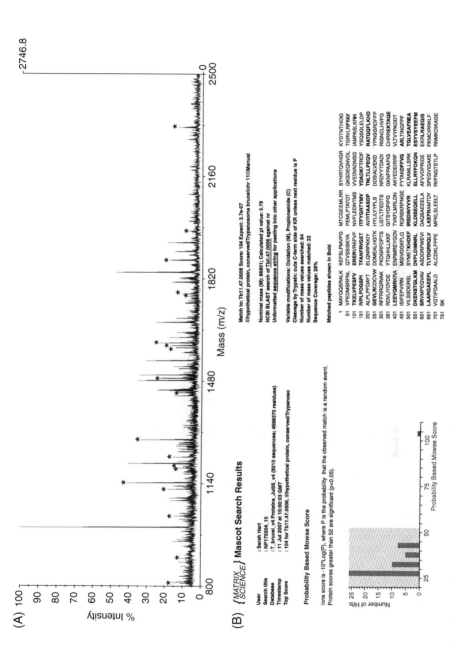

Figure 1. Peptide mass fingerprinting using conventional mass spectrometry. (A) MALDI-TOF analysis was performed upon a tryptic digest of a Coomassie-stained band from a 1-D gel separation of Typanosoma brucei flagella. (B) Peptide mass values were submitted for Mascot searching, which resulted in the recognition of a single component. Mass values that were assigned to a putative peptide are marked with an asterisk on the spectrum in (A). A number of prominent peaks were unassigned, suggesting the presence of peptides from more than one protein component within this mixture.

Table 1. Commonly used search tools for assignment of peptide mass spectra generated by conventional and tandem mass spectrometry analyses.

Name, url (if web-based), licensing arrangements for local installation	PMF, MS/MS or both?	Advantages	Disadvantages	References
Mascot http://www.matrixscience.com/ Local licensing possible	Both	Widely accepted as industry standard. Iterative searching ('error-tolerant') on limited database entries possible to increase surety of match. Mass error plots of results enable identification of matches which are poor based on mass accuracy of peptides (PMF or MS/MS) and/or product ions (MS/MS) False positive rate calculation incorporated in Mascot 2.2	Local license essential for searching large datasets and adding custom databases, modifications or enzymes. Algorithm unpublished. Default threshold for confident matching p<0.05	(Perkins, Pappin *et al.* 1999; Creasy and Cottrell 2002)
Phenyx http://www.phenyx-ms.com/ Local licensing possible	MS/MS	Iterative searching (resubmit with refinements) on limited database entries possible to increase surety of match. Can define and alter 'acceptance parameters', thresholds for confident match. Can add custom modifications and enzymes	Not very widely used	(Heller, Ye *et al.* 2005)
Protein Prospector http://prospector.ucsf.edu Local licensing possible	Both	Wide variety of tools for PMF, MS/MS, *de novo* sequencing and confirmation of potential hits available	Export of results from web-based server in standards-compliant format difficult	(Clauser, Baker *et al.* 1999)
Sequest Local licensing essential	MS/MS	Widely *used*. Intensity-based matching of product ion spectra. Can add custom sequence databases, modifications and enzymes	Tied to specific instrument vendor and data formats. Original algorithm published, but substantially modified for commercial application. Search speeds typically slow. Databases must be indexed for any change to enzyme or modifications	(Eng, McCormack *et al.* 1994)
X! Tandem http://h.thegpm.org/tandem/ thegpm_tandem.html	MS/MS	Facile comparison with previously identified datasets via gpmDB. Iterative searching ('error-tolerant') on limited database entries possible to increase surety of match. Thresholds for confidence of match can be set. Can add custom modifications and enzymes	Limited organisms, annotation of data minimal. Scoring scheme not particularly clear	(Craig and Beavis, 2004; Craig Cortens *et al.* 2006)

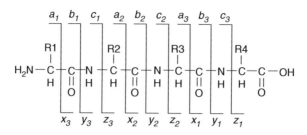

Figure 2. *Peptide fragmentation by tandem mass spectrometry: standard nomenclature. The exact masses of the different fragments will depend on the protonation state of the precursor and product ions.*

the development of a number of commercial and open-source search algorithms which compare theoretical peptide fragmentation patterns with experimentally derived data (Clauser *et al.*, 1999; Craig and Beavis, 2004; Craig *et al.*, 2006; Eng *et al.*, 1994; Heller *et al.*, 2005; Perkins *et al.*, 1999; see *Table 1*). Partial sequence data derived from CAD of peptide precursor ions is used in database searching strategies to infer peptide identity and allow protein recognition by comparison with sequence databases (see *Figure 3*). Such analyses rely heavily on high-quality sequence data-bases, which only exist for a limited subset of organisms. Where such information does not exist, peptide homology searching to proteolytic fragments from related organisms with sequenced genomes may be used (Shevchenko *et al.*, 2001); however, in many cases homology is not sufficient to infer protein recognition, and *de novo* sequence inference from product ion spectra is the only method available (Liao *et al.*, 2007; Waridel *et al.*, 2007).

Traditionally, the use of tandem mass spectrometry in peptide sequence analysis has been based upon CAD methods almost without exception. More recently, new methods of ion activation have been developed and applied to the analysis of peptides and proteins. Electron capture dissociation (Zubarev *et al.*, 1998, 2000) and latterly electron transfer dissociation (ETD; Coon *et al.*, 2005; Mikesh *et al.*, 2006; Syka *et al.*, 2004) methods initiate radical-induced fragmentation of polypeptides via an alterna-tive pathway, inducing cleavage of the $N–C\alpha$ bond. This cleavage generates c- and z-series ions derived from peptide N- and C-termini respectively (*Figure 2*). Such methods have demonstrated high utility as an alternative to CAD, particularly in the analysis of post-translationally modified peptides (Chi *et al.*, 2007; Schroeder *et al.*, 2005; Syka *et al.*, 2004) and intact proteins (Zabrouskov *et al.*, 2005; Zubarev *et al.*, 2000), and also show general utility in fragmentation of peptide ions which are intractable by CAD (see *Figure 4*). The exploitation of ETD may presage a greater reliance on the use of bimolecular gas-phase ion chemistry for the characterization of polypeptides using tandem mass spectrometry.

Standardization in reporting of proteomic datasets aims to integrate data from different laboratories, running myriad mass spectrometric platforms, and report and store data in a meaningful and robust manner, which can be subjected to facile com-parison and future re-interrogation. Amongst the major initiatives in proteome stan-dards development is the HUPO Proteome Standards Initiative (www.psidev.info)

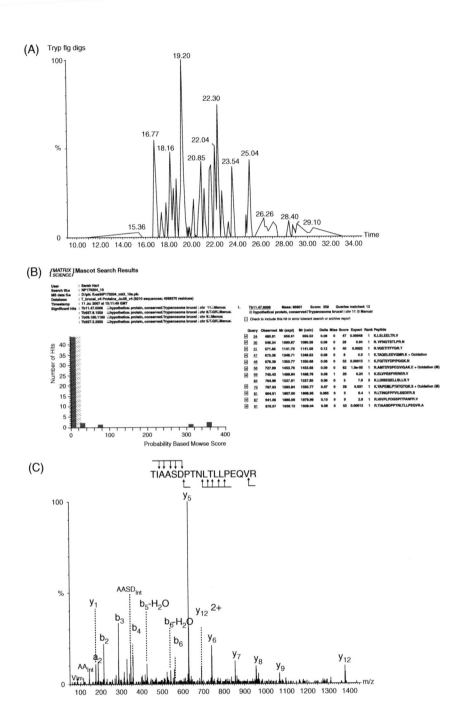

Figure 3. *Peptide sequence analysis by tandem mass spectrometry. (A) Base peak ion chromatogram of the tryptic digest as in Figure 1. Multiply charged precursor ions were selected for tandem mass spectrometry in a data-directed experiment. Product ion mass spectra were converted to text files for database searching, (B), resulting in the recognition of sequences originating from four protein components within this mixture, including the same protein postulated by the data in Figure 1. (C) Typical product ion spectrum of a peptide, sequence TIAASDPTNLTLLPEQVR, showing strong proline-directed cleavage (y$_5$). Peptide fragmentation nomenclature is as shown in Figure 2.*

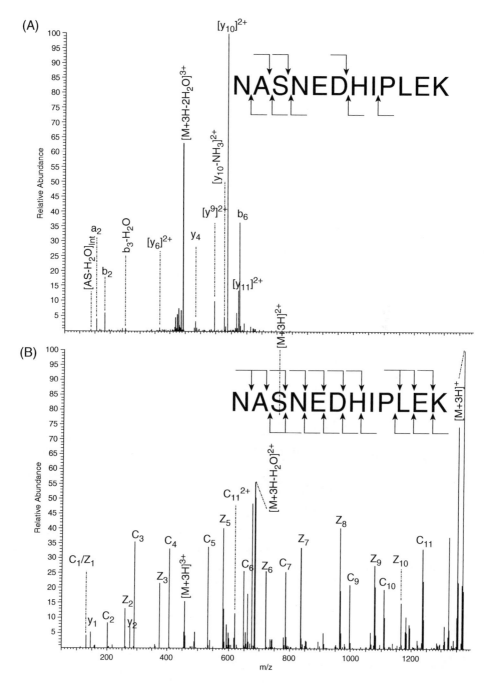

Figure 4. Comparison of product ion spectra generated by collision activated dissociation and electron transfer dissociation. (A) Product ion spectrum generated by CAD from a triply protonated dodecapeptide, sequence NASNEDHIPLEK. Limited sequence information was derived from this product ion analysis. (B) Product ion spectrum generated by ETD from the same dodecapeptide precursor ion. Full sequence information with the exception of the expected lack of cleavage at the imino acid proline was derived from this product ion analysis.

(Hermjakob, 2006), which seeks to define standards for proteomic data definition by defining controlled ontological vocabularies and XML schema for data reporting. This effort takes into account many of the consortium- and journal-specific guidelines that have previously been developed (Bradshaw *et al.*, 2006; Pedrioli *et al.*, 2004; Taylor *et al.*, 2003; Wilkins *et al.*, 2006). Various data repositories for the storage and sharing of proteomic data have also been developed (Coon *et al.*, 2005; Desiere *et al.*, 2006; Jones *et al.*, 2006; Orchard *et al.*, 2007). Ultimately, the utility for these resources will be in their ability to enable independent verification of results collected in multiple laboratories, and in the use of the data contained therein to derive and prove novel data-driven hypotheses (Shadforth and Bessant, 2006).

Recent effort has focused upon statistical validation of database search results, and recognition of the number of both false positive and false negative 'hits' within search returns. Currently, interest involves minimizing the number of false positive hits via careful treatment of data, whilst allowing the maximum number of true positive matches to be made (Elias and Gygi, 2007; Peng *et al.*, 2003a). Current best practice within shotgun proteomics involves the independent recognition of two or more non-redundant peptides whose cumulative score is above a stated threshold, the use of which correlates with a specific false positive rate, generated by searching against a decoy database containing identical amino acid composition. Other ways to increase the confidence of matched tandem mass spectrometric data are the use of multiple independent search programs (see *Table 1*), each of which uses different methods to match sequence data to mass spectrometric profiles; this combined approach can improve the statistical confidence in database search results (Keller *et al.*, 2002; Nesvizhskii *et al.*, 2003). Such strategies are readily built into informatic workflows and are rapidly becoming a standard feature of proteomic pipelines (Keller *et al.*, 2005; Shadforth *et al.*, 2006).

2.4 *Characterization of post-translational modifications*

Proteome-wide characterization of post-translational modifications is of significant interest within proteomics and mass spectrometry because of the high biological importance of many of these modifiers. The ability of standard proteomic methods to achieve comprehensive analysis of post-translational modifications is generally poor as a result of a number of factors, including the low abundance and stoichiometry of the modified proteins/peptides, poor efficiency of generation of gas-phase ions from modified peptides and hence poor detection efficiency (Mann *et al.*, 2002; Mann and Jensen, 2003). As a result of these factors, the majority of approaches to characterizing post-translational modifications are targeted ones, either using specific sample preparation methods or directed mass spectrometric experiments. Applications to many post-translational modifications have been developed, such as acetylation (Xie *et al.*, 2007), *O*-sulphation (Yu *et al.*, 2007), *N*- (Kaji *et al.*, 2003; Krokhin *et al.*, 2004) and *O*-glycosylation (Chalkley and Burlingame, 2003; Harvey, 2005; Vosseller *et al.*, 2006) and ubiquitination (Kirkpatrick *et al.*, 2005a, Peng *et al.*, 2003b). For the sake of brevity, however, only the most common application, phosphorylation site analysis, will be described here. Major effort has been devoted to the analysis of phosphorylation over the last decade, with several examples of both selective enrichment and targeted mass spectrometry being developed.

Isolation or enrichment of post-translationally-modified proteins and/or peptides is a common means to improve the detection efficiency of a specific class of analytes, purely by removal of other confounding analytes from the sample mixture. Selective isolation has been achieved with intact phosphoproteins from a background of non-phosphorylated proteins (Collins *et al.*, 2005a, 2005b; Guerrera *et al.*, 2005; Makrantoni *et al.*, 2005); this step does not, of course, resolve the issue of predominant non-phosphorylated peptides following subsequent enzymatic hydrolysis. Strategies which involve selective isolation of phosphorylated peptides have therefore been widely investigated. Two major methods have emerged, both based upon interaction of phosphorylated peptides with metal ions upon a stationary phase support. The first of these, immobilized metal ion affinity chromatography (IMAC), has been used in a number of phosphoprotein (Barnouin *et al.*, 2005; Haydon *et al.*, 2003; Neville *et al.*, 1997; Stensballe and Jensen, 2004; Thompson *et al.*, 2003) and phosphoproteome (Collins *et al.*, 2005b; Ficarro *et al.*, 2002, 2005; Nuhse *et al.*, 2003; Salomon *et al.*, 2003) investigations. This method, which relies on electrostatic interaction of metal with phosphate groups, shows high affinity for phosphopeptides, but relatively low specificity, with strong retention of non-phosphorylated peptides via acidic amino acid side chains being widely reported (Hart *et al.*, 2002; Haydon *et al.*, 2003; Neville *et al.*, 1997). Prior formation of peptide methyl esters to prevent this interaction has been reported as a means to improve specificity (Ficarro *et al.*, 2002). The second major method, retention of peptides on titanium dioxide (TiO_2) beads, has been reported to show higher specificity for phosphopeptides versus acidic components (Larsen *et al.*, 2005; Mazanek *et al.*, 2007; Pinkse *et al.*, 2004; Thingholm *et al.*, 2006); the two methods have recently been reported to have overlapping but complementary phosphopeptide retention characteristics by Bodenmiller *et al.* (2007). In the same publication, a third method, using phosphoramidate chemistry (first reported by Zhou *et al.*, 2001), was found also to have an overlapping but complementary retention profile when compared with the metal chelate-base methods Thus a comprehensive analysis of the phosphoproteome may require multiple stationary phase and derivatization strategies to be combined.

Mass spectrometric methods for selective detection and determination of phosphorylated peptides have also proved fruitful. These include methods which search for specific fragment ions which indicate phosphorylation (Annan *et al.*, 2001; Steen *et al.*, 2001), data-directed analyses where the prominent neutral loss of phosphate triggers a further stage of tandem mass spectrometry with CAD of the product ion of neutral loss (NL-triggered MS^3; Le Blanc *et al.*, 2003), and predictive methods for individual proteins, where selected peptide/precursor ion transitions are searched for to provide evidence for specific phosphorylation sites (Griffiths *et al.*, 2007; Unwin *et al.*, 2005). The majority of these methods rely on the fact that upon collisional activation, phosphoserine- and phosphothreonine-containing protonated peptides readily undergo neutral loss of phosphoric acid (see Mann *et al.*, 2002 for a review). This facile neutral loss means that identification of the precise residue bearing the modification is often difficult, since the majority of the ion current is directed along this non-specific channel. Alternative methods such as ECD and ETD may show better sequence coverage for phosphopeptides, because the mechanism and fragmentations induced by electron-mediated cleavage are very different. Phosphopeptides and other post-translationally modified peptides have proved readily tractable using

ECD and ETD (Chalmers *et al.*, 2004; Coon *et al.*, 2005; Stensballe *et al.*, 2000; Syka *et al.*, 2004).

3 Quantitative proteome analyses

Quantitative proteomics as a field encompasses a number of different types of study, including relative quantification based upon comparison (for example) of two samples grown under different conditions, and absolute quantification to determine the amount of a protein in a defined sample. Additional quantitative studies may seek to define the rates of protein synthesis and degradation. A common principle in the majority of these methods is that of surrogacy – peptides are used as surrogate analytes for their protein progenitors, and the molar concentration of peptide is assumed to be that of the protein. This section defines some of the pertinent issues in this field as a whole, and draws the reader's attention to key references drawn from the extensive literature.

3.1 *Relative quantification strategies*

The majority of relative quantification approaches share a common principle, that of differential labelling of samples. Differential labelling of proteins using fluorophores bearing distinct absorption/emission wavelengths but equivalent molecular weights was first used in relative quantification of 2-D gel-separated proteins by Unlu and colleagues (1997). This method, termed difference gel electrophoresis (or DiGE) was rapidly adopted by a number of groups to perform relative quantification experiments on highly complex samples (Gharbi *et al.*, 2002; Kernec *et al.*, 2001; Lilley and Friedman, 2004; Swatton *et al.*, 2004; Viswanathan *et al.*, 2006). The initial application using minimal labelling with *N*-hydroxy succinimide ester-based dyes (see Viswanathan *et al.*, 2006 for a protocol), which react with lysine side chains, was rapidly realized to have a number of drawbacks, including differential migration of labelled proteins from their bulk, unlabelled counterparts, necessitating post-staining for protein visualization, and feature picking for analysis. Necessarily this process significantly increased the timescales for such analyses. Saturation labelling of cysteine residues with maleimide-linked dyes was therefore introduced as a means to overcome these shortcomings and improve detection sensitivity (Lilley and Friedman, 2004; Shaw *et al.*, 2003). Ultimately, however, the utility of DiGE suffers as a result of the limited dynamic range, loading and resolving capabilities of the 2-D gel electrophoresis method itself, as described above (Section 2.1).

Improved capability of mass spectrometric instrumentation and new strategies for stable isotope incorporation mean that stable isotope dilution methods now predominate in quantitative proteome analyses. The approach may involve metabolic incorporation of a label during tissue culture or growth of model organisms, or the labelling of intact proteins or their proteolytic peptides. *Figure 5* summarizes the different approaches. Each approach has advantages and disadvantages, though in general it is beneficial to introduce differential isotope labelling as early in the workflow as possible. In all cases, quantification is based on the measurement of relative signal intensities for the isotopic variants of the same peptide.

Metabolic incorporation of stable isotopes into proteins can be performed in a variety of ways. For instance, common precursors of amino acids can be used to perform

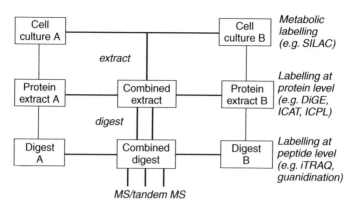

Figure 5. *Schematic diagram of conventional workflows in quantitative proteome analyses. Quantitative proteome analyses based upon labelling strategies can be categorized according to the stage at which the label is incorporated and parallel samples mixed together for relative .determination of analyte concentration. The earlier that samples are mixed together for analysis, the lower the anticipated error of the resultant data. Prominent examples of each strategy, referred to within the text, are listed here.*

global labelling of proteins. Strategies using [^{13}C]glucose as the sole carbon source (Cargile *et al.*, 2004) or [^{15}N]H$_4$Cl as the sole nitrogen source (Krijgsveld *et al.*, 2003; Smith *et al.*, 2002) have been developed. A disadvantage of these methods is that proteolytic peptides necessarily contain differing numbers of carbon and nitrogen atoms, so that the mass differences between naturally occurring and stable isotopic variants are also variable, complicating the task of identifying pairs. Stable isotope incorporation via the use of amino acids is an alternative approach; judicious selection of labelled amino acid is, however, essential (Beynon and Pratt, 2005). Stable isotope incorporation via the use of amino acids results in predictable mass differences, dependent solely upon the number of labelled amino acids within a single peptide. The introduction of stable isotope labelling with essential amino acids in cultured cells (also known as SILAC; Ong and Mann, 2006; Ong *et al.*, 2002) has enabled a number of applications in the analysis of relative protein expression to be performed (Amanchy *et al.*, 2005; Foster *et al.*, 2006; Gruhler *et al.*, 2005; Ong *et al.*, 2003; Thiede *et al.*, 2006; Zhang *et al.*, 2006). The introduction of stable isotopes into proteins during their synthesis can be extended from incorporation of stable isotopes in cell culture media (Ong and Mann, 2007), via feeding insects on bacteria or yeast which have been grown in isotope-labelled media (Krijgsveld *et al.*, 2003), to feeding animals on a stable isotope-incorporating diet (Doherty *et al.*, 2004a, 2004b; Pratt *et al.*, 2002). A difficulty with these last methods is that tolerance manipulation of the diet of an organism is far from facile, particularly with higher organisms; in addition, the relative isotope abundance of the actual precursor pool must be determined (Doherty *et al.*, 2004a).

Post-synthetic incorporation of stable isotopes into intact proteins enables labelling of proteins from any source, including clinical samples. Strategies have been developed based upon cysteine derivatization with stable isotope-labelled affinity tags (Gygi *et al.*, 1999; Olsen *et al.*, 2004) or formation of *N*-hydroxy succinimidyl

(A)

(B)

Derivative 'flavour'	m/z reporter ion	Tag heavy isotope content	Balance heavy isotope content
A	114	+0	+3
B	115	+1	+2
C	116	+2	+1
D	117	+3	+0

(C)

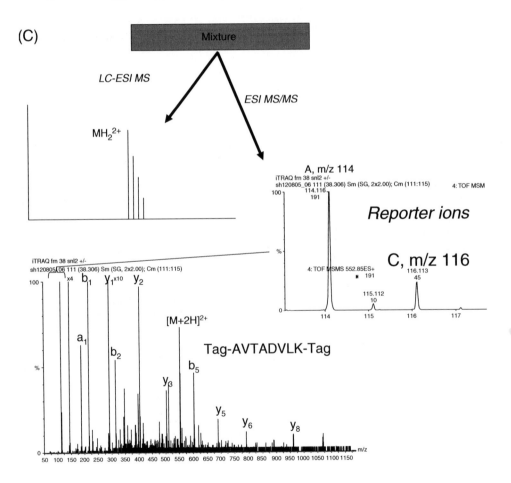

Figure 6. Schematic of the principles used in quantitative proteome analyses with isobaric tags for relative and absolute quantification. (A) Four distinct tags are used in strategies involving isobaric tags for relative and absolute quantification (iTRAQ). Tagging reagents consist of three elements; a reactive N-hydroxy succinimide that reacts with primary amine groups upon peptides, a reporter ('tag') that provides the relative quantification in tandem mass spectrometry, and a balance that makes the precursor ions from peptides labelled with all four variants isobaric. (B) Table illustrating the number of 'heavy' stable isotope-labelled atoms in each reagent within the tag and balance moieties. (C) Typical iTRAQ experiment; peptides from different cell preparations which are to be compared each undergo labelling with a specific isobaric tag. They are then mixed together and subjected to LC-tandem MS, during which selected peptides are fragmented by CAD. Product ion spectra contain two sets of information: the sequence-related b- and y-series ions, which are used to allow recognition of the peptide sequence, and reporter ion intensities, which are used to determine the relative abundances of the peptides and therefore proteins. In the case shown, two distinct iTRAQ tags were used to label peptides from proteolytic digests from T. brucei flagella; the reporter ion intensities indicated a four-fold decrease in the expression level of adenylate kinase.

esters at lysine side chains (Schmidt *et al.*, 2005; Wiese *et al.*, 2007). These methods have the advantage that a label is introduced and samples are mixed prior to proteolysis, minimizing errors induced by differential efficiencies of enzymatic digestion. The achievement of efficient derivatization at the protein level may, however, be challenging.

Incorporation of stable isotopes during (Reynolds *et al.*, 2002) or post-digestion (DeSouza *et al.*, 2005; Unwin *et al.*, 2006) is technically simple, but as a result of the late stage of sample mixing can lead to a high degree of analytical variability (Chong *et al.*, 2006). One method which is currently extremely popular is that of isobaric tagging for relative and absolute quantification (iTRAQ™, see *Figure 6*). In this method, isobaric derivatives bearing 'tag' and 'balance' regions label peptide amine groups. Upon CAD during tandem MS, the tag itself fragments to give a 'reporter' ion, the abundance of which correlates with the concentration of the precursor peptide population. These reagents are multiplexable (currently four variants are commercially available, with an 8-plex kit being imminently released at time of writing), which means that multiple samples can readily be compared in a single experiment. High selectivity results from the use of tandem mass spectrometry of a defined precursor ion population for quantification. We recently reported an inexpensive approach to relative quantification, which can also benefit from the selectivity of tandem MS analysis. The method is based upon derivatization of lysine side chains in peptides using O-methyl isourea (Warwood *et al.*, 2006). In this strategy, two samples are derivatized, one with [^{15}N]-containing reagent and the other with the naturally occurring isotope. Product ion analysis is performed using a sufficiently wide precursor ion window to allow simultaneous transmission of both isotopic variants; y-series product ions from lysine containing peptides are then used to determine the relative quantity of peptide in the two samples.

Stable isotope incorporation strategies generally require that correction be made for the incomplete isotopic purity of the labelled material. Experimental variability arises (and must be assessed) when isotopic differentiation is deferred until isolation of protein or peptide fractions. A recent approach for relative quantification without the use of stable isotopes has emerged, termed label-free quantification (Wiener *et al.*, 2004). These methods rely upon highly reproducible sample preparation, HPLC and mass spectrometry methods, with *post hoc* chromatographic realignment for identification of pertinent varying factors between different samples (Bellew *et al.*, 2006; Silva *et al.*, 2006). Such methods are appealing in avoiding the need for stable isotope labelling.

3.2 *Absolute quantification*

Absolute quantification provides concentrations of individual proteins in molar terms or as copy numbers per cell. An extension of the label-free quantification approach described in the last section is reported to provide approximate absolute quantification, based on the (perhaps surprising) observation that the average individual mass spectrometric signal intensity of the three most abundant tryptic peptides is directly related to protein amount, regardless of the individual protein (Bellew *et al.*, 2006). Precise and accurate absolute quantification, however, is likely generally to require application of the same analytical principles – of surrogacy and isotope dilution – that were discussed above in the context of relative quantification. The requirement is for the availability in known amounts of one or more isotopically labelled tryptic peptide for every protein to be quantified.

Two general approaches have been adopted. The first relies on chemical synthesis of stable isotope-containing peptides, and has been termed AQUA (Gerber *et al.*, 2003; Kirkpatrick *et al.*, 2005b). This method, whilst reproducible and generating high yields of peptide, must be executed on a peptide-by-peptide basis. This is therefore costly and requires characterization of purity and concentration of each individual peptide. One significant advantage of this approach is that for phosphopeptides, phosphorylated amino acids can be incorporated readily, allowing absolute quantification of phosphorylated peptides (Gerber *et al.*, 2007; Mayya *et al.*, 2006).

The second approach uses the natural machinery of the cell to produce quantification standards, and was developed by Beynon *et al.* (2005) in collaboration with this laboratory. This method, termed QconCAT (QCAT in the earliest paper), uses exogenous protein expression from plasmids introduced to bacterial cells to 'synthesize' stable isotope-containing artificial proteins, comprising a concatenation of tryptic peptides selected from the sequences of proteins which are to be quantified. Quantification of the QconCAT protein is performed on each purified batch. In the first application of this method, 20 tryptic peptides serving as signatures for 20 proteins were produced (Beynon *et al.*, 2005); more recent studies have shown that larger QconCATs, representing higher numbers of tryptic peptides, are readily expressed. Since plasmid stocks, competent cells and stable isotope-containing media (typically $[^{13}C_6]$-arginine and -lysine, ensuring that all tryptic peptides contain stable isotope label; Pratt *et al.*, 2006; Rivers *et al.*, 2007) are the main requirements, QconCAT protein stocks are theoretically inexhaustible, and the costs of generation of labelled peptide stocks via this route can be significantly lower than by chemical synthesis. Accurate quantification using the QconCAT method requires efficient tryptic cleavage of both endogenous proteins and the QconCAT standard. Kito *et al.* (2007) addressed this issue by incorporating flanking residues around tryptic peptides in artificial protein sequences. However, substantiation of the quantitative approach still requires assessment of the validity of the assumption of complete digestion.

Two further issues should be addressed in considering quantitative proteomics methods, whether relative or absolute. The first is that the use of one or more tryptic peptides as 'signature' or 'proteotypic' peptides for the protein of interest necessarily means that the analytical method is blind to structural variations elsewhere in the protein. Thus, these methods quantify all proteins incorporating the (wisely or not) selected tryptic peptide sequence. Second, many applications to date have made use of mass spectrometric analytical modes that have a low duty cycle (meaning that a low proportion of the total analysis time is devoted to detection of signals associated with individual peptides). Enhanced duty cycles and high selectivities of analysis are obtained using selected reaction monitoring during LC-tandem MS and such methods are likely to assume increasing importance.

4 The kinetics of the proteome

Quantitative determinations of the components of the proteome provide the net results of processes of synthesis, modification and degradation. A full, 'multi-dimensional' definition of the proteome therefore requires quantitative determination of these dynamic processes. This poses new analytical challenges that have only begun to be addressed but which must be met for a full contribution of proteomics to systems biology; studies indicative of the possibilities are summarized briefly here.

4.1 *Protein turnover*

The rate of protein turnover is regulated by transcription of DNA, translation into protein, processing, modification and degradation, and therefore illustrates the dynamic interaction between '-omes' which will need to be defined to enable an integrated systems description of cellular behaviour (Doherty and Beynon, 2006). Within unicellular organisms, such as yeast or bacteria, and in mammalian cell cultures, it is possible to produce fully stable isotope-labelled cell lines using a medium containing one or more labelled amino acids. A 'pulse chase' of unlabelled amino acid then enables the measurement of the rates of turnover of individual proteins via determination of the variation of isotopic composition of proteins over time (Cargile *et al.*, 2004; Pratt *et al.*, 2002). This approach can be extended to the analysis of the dynamic proteome of higher organisms via incorporation of a labelled amino acid into feedstuffs (Doherty *et al.*, 2004a) or by catheterization (Papageorgopoulos *et al.*, 1999, 2002; Vogt *et al.*, 2005) for labelling of animals. In such experiments, the direct or indirect determination of relative isotope abundance in the precursor pool is required.

4.2 *Kinetics of phosphorylation*

Post-translational modification of proteins is amongst the most tightly regulated processes within the cell, dysregulation of which is closely implicated in the pathology of cancer and other diseases (Hanahan and Weinberg, 2000; Hunter, 1998). Investigation of the dynamic nature of phosphorylation and other post-translational modifications using quantitative proteomic methods is therefore a fertile field for investigation. Recent work by Mann and co-workers investigated the dynamic nature of phosphorylation using SILAC labelling methods and receptor-ligand stimulation over defined time courses (Dengjel et al., 2007), to define protein function within receptor-mediated signalling cascades (Olsen et al., 2006). The analytical strategy involved hierarchical clustering of phosphorylated peptides isolated by TiO_2 and analysis by neutral loss-triggered MS^3. This is an area where the convergence of proteomic strategies for quantitative analysis of phosphorylation site status promises numerous additional applications within the near future.

5 Concluding remarks and future challenges

In comparison with an organism's genome, the proteome represents an enormously complex analytical target. The present literature, however extensive, represents no more than a scratching of the surface. Most proteomic studies concern the most abundant proteins only, take little account of the full extent of post-translational modification, are rarely quantitative, and disregard the dynamics of the processes that yield a net proteome. This is not to deny the value of previous work, nor to disregard the elegance of the analytical techniques that have enabled progress thus far. Nevertheless, the remaining challenges are substantial.

First amongst these is the sensitivity of analysis. In general, proteome analyses are based on numbers of cells well in excess of 10^5; single cell proteomics remains an important but distant objective whose achievement will require attention to multiple facets of existing analytical procedures. The related issue of dynamic range is also critical; achievement of a dynamic range of more than 10^3 is presently rare in a single analytical protocol. Great ingenuity has been applied to the characterization of post-translational modifications, but the variety of such structural changes and the low stoichiometry in which they frequently occur mean that much remains to be done before either qualitative or quantitative determinations are achieved. The importance of quantitative analyses (both relative and absolute) of proteins is now widely appreciated but both accuracy and precision are commonly modest, reflecting inadequate attention to optimization of the analytical modes employed. Finally, the full value of proteomics data in a quantitative description of biological systems will only be realized when the dynamics of the proteome are properly defined.

References

Amanchy, R., Kalume, D.E. & Pandey, A. (2005) Stable isotope labeling with amino acids in cell culture (SILAC) for studying dynamics of protein abundance and posttranslational modifications. *Sci. STKE*: p l2.

Annan, R.S., Huddleston, M.J., Verma, R., Deshaies, R.J. & Carr, S.A. (2001) A multidimensional electrospray MS-based approach to phosphopeptide mapping. *Anal. Chem.* **73**: 393–404.

Barnouin, K.N., Hart, S.R., Thompson, A.J., Okuyama, M., Waterfield, M. & Cramer, R. (2005) Enhanced phosphopeptide isolation by Fe(III)-IMAC using 1,1,1,3,3,3-hexafluoroisopropanol. *Proteomics* **5**: 4376–4388.

Bellew, M., Coram, M., Fitzgibbon, M. *et al.* (2006) A suite of algorithms for the comprehensive analysis of complex protein mixtures using high-resolution LC-MS. *Bioinformatics* **22:** 1902–1909.

Beynon, R.J. & Pratt, J.M. (2005) Metabolic labeling of proteins for proteomics. *Mol. Cell Proteomics* **4:** 857–872.

Beynon, R.J., Doherty, M.K., Pratt, J.M. & Gaskell, S.J. (2005) Multiplexed absolute quantification in proteomics using artificial QCAT proteins of concatenated signature peptides. *Nat. Methods* **2:** 587–589.

Biemann, K. (1988) Contributions of mass spectrometry to peptide and protein structure. *Biomed. Environ. Mass Spectrom.* **16:** 99–111.

Blencowe, B.J. (2006) Alternative splicing: new insights from global analyses. *Cell* **126:** 37–47.

Blonder, J., Conrads, T.P. & Veenstra, T.D. (2004) Characterization and quantitation of membrane proteomes using multidimensional MS-based proteomic technologies. *Expert Rev. Proteomics* **1:** 153–163.

Blum, H.B. & Gross, H.J. (1987) Improved silver staining of plant proteins, RNA and DNA in polyacrylamide gels. *Electrophoresis* **8:** 93–99.

Bodenmiller, B., Mueller, L.N., Mueller, M., Domon, B. & Aebersold, R. (2007) Reproducible isolation of distinct, overlapping segments of the phosphoproteome. *Nat. Methods* **4:** 231–237.

Bradshaw, R.A., Burlingame, A.L., Carr, S. & Aebersold, R. (2006) Reporting protein identification data: the next generation of guidelines. *Mol. Cell Proteomics* **5:** 787–788.

Brancia, F.L., Butt, A., Beynon, R.J., Hubbard, S.J., Gaskell, S.J. & Oliver, S.G. (2001) A combination of chemical derivatisation and improved bioinformatic tools optimises protein identification for proteomics. *Electrophoresis* **22:** 552–559.

Brett, D., Pospisil, H., Valcarcel, J., Reich, J. & Bork, P. (2002) Alternative splicing and genome complexity. *Nat. Genet.* **30:** 29–30.

Broadhead, R., Dawe, H.R., Farr, H. *et al.* (2006) Flagellar motility is required for the viability of the bloodstream trypanosome. *Nature* **440:** 224–227.

Cargile, B.J., Bundy, J.L., Grunden, A.M. & Stephenson, J.L., Jr. (2004) Synthesis/degradation ratio mass spectrometry for measuring relative dynamic protein turnover. *Anal. Chem.* **76:** 86–97.

Carrette, O., Burkhard, P.R., Sanchez, J.C. & Hochstrasser, D.F. (2006) State-of-the-art two-dimensional gel electrophoresis: a key tool of proteomics research. *Nat. Protoc.* **1:** 812–823.

Chalkley, R.J. & Burlingame, A.L. (2003) Identification of novel sites of O-N-acetylglucosamine modification of serum response factor using quadrupole time-of-flight mass spectrometry. *Mol. Cell Proteomics* **2:** 182–190.

Chalmers, M.J., Hakansson, K., Johnson, R., Smith, R., Shen, J., Emmett, M.R. & Marshall, A.G. (2004) Protein kinase A phosphorylation characterized by tandem Fourier transform ion cyclotron resonance mass spectrometry. *Proteomics* **4:** 970–981.

Chi, A., Huttenhower, C., Geer, L.Y. *et al.* (2007) Analysis of phosphorylation sites on proteins from Saccharomyces cerevisiae by electron transfer dissociation (ETD) mass spectrometry. *Proc. Natl Acad. Sci. USA* **104:** 2193–2198.

Chong, P.K., Gan, C.S., Pham, T.K. & Wright, P.C. (2006) Isobaric tags for relative and absolute quantitation (iTRAQ) reproducibility: Implication of multiple injections. *J. Proteome Res.* **5**: 1232–1240.

Clauser, K.R., Baker, P. & Burlingame, A.L. (1999) Role of accurate mass measurement (+/- 10 ppm) in protein identification strategies employing MS or MS/MS and database searching. *Anal. Chem.* **71**: 2871–2882.

Cohen, P. (2000) The regulation of protein function by multisite phosphorylation— a 25 year update. *Trends Biochem. Sci.* **25**: 596–601.

Collins, M.O., Yu, L., Coba, M.P., Husi, H., Campuzano, I., Blackstock, W.P., Choudhary, J.S. & Grant, S.G. (2005a) Proteomic analysis of in vivo phosphorylated synaptic proteins. *J. Biol. Chem.* **280**: 5972–5982.

Collins, M.O., Yu, L., Husi, H., Blackstock, W.P., Choudhary, J.S. & Grant, S.G. (2005b) Robust enrichment of phosphorylated species in complex mixtures by sequential protein and peptide metal-affinity chromatography and analysis by tandem mass spectrometry. *Sci. STKE*: p l6.

Coon, J.J., Ueberheide, B., Syka, J.E., Dryhurst, D.D., Ausio, J., Shabanowitz, J. & Hunt, D.F. (2005) Protein identification using sequential ion/ion reactions and tandem mass spectrometry. *Proc. Natl Acad. Sci. USA* **102**: 9463–9468.

Craig, R. & Beavis, R.C. (2004) TANDEM: matching proteins with tandem mass spectra. *Bioinformatics* **20**: 1466–1467.

Craig, R., Cortens, J.C., Fenyo, D. & Beavis, R.C. (2006) Using annotated peptide mass spectrum libraries for protein identification. *J. Proteome Res.* **5**: 1843-1849.

Creasy, D.M. & Cottrell, J.S. (2002) Error tolerant searching of uninterpreted tandem mass spectrometry data. *Proteomics* **2**: 1426–1434.

Dengjel, J., Akimov, V., Olsen, J.V., Bunkenborg, J., Mann, M., Blagoev, B. & Andersen, J.S. (2007) Quantitative proteomic assessment of very early cellular signaling events. *Nat. Biotechnol.* **25**: 566–568.

Desiere, F., Deutsch, E.W., King, N.L. *et al.* (2006) The PeptideAtlas project. *Nucleic Acids Res.* **34**: D655–658.

DeSouza, L., Diehl, G., Rodrigues, M.J., Guo, J., Romaschin, A.D., Colgan, T.J. & Siu, K.W. (2005) Search for cancer markers from endometrial tissues using differentially labeled tags iTRAQ and cICAT with multidimensional liquid chromatography and tandem mass spectrometry. *J. Proteome Res.* **4**: 377–386.

Doherty, M.K. & Beynon, R.J. (2006) Protein turnover on the scale of the proteome. *Expert Rev. Proteomics* **3**: 97–110.

Doherty, M.K., McClean, L., Edwards, I., McCormack, H., McTeir, L., Whitehead, C., Gaskell, S.J. & Beynon, R.J. (2004a) Protein turnover in chicken skeletal muscle: understanding protein dynamics on a proteome-wide scale. *Br. Poult. Sci.* **45 Suppl. 1**: S27–28.

Doherty, M.K., McLean, L., Hayter, J.R., Pratt, J.M., Robertson, D.H., El-Shafei, A., Gaskell, S.J. & Beynon, R.J. (2004b) The proteome of chicken skeletal muscle: changes in soluble protein expression during growth in a layer strain. *Proteomics* **4**: 2082–2093.

Elias, J.E. & Gygi, S.P. (2007) Target-decoy search strategy for increased confidence in large-scale protein identifications by mass spectrometry. *Nat. Methods* **4**: 207–214.

Eng, J.K., McCormack, A.L. & Yates, I.J.R. (1994) An approach to correlate tandem mass spectral data of peptides with amino acid sequences in a protein database. *J. Am. Soc. Mass Spectrom.* **5**: 976–989.

Ficarro, S.B., McCleland, M.L., Stukenberg, P.T., Burke, D.J., Ross, M.M., Shabanowitz, J., Hunt, D.F. & White, F.M. (2002) Phosphoproteome analysis by mass spectrometry and its application to Saccharomyces cerevisiae. *Nat. Biotechnol.* **20**: 301–305.

Ficarro, S.B., Salomon, A.R., Brill, L.M., Mason, D.E., Stettler-Gill, M., Brock, A. & Peters, E.C. (2005) Automated immobilized metal affinity chromatography/ nano-liquid chromatography/electrospray ionization mass spectrometry platform for profiling protein phosphorylation sites. *Rapid Commun. Mass Spectrom.* **19**: 57–71.

Foster, L.J., Rudich, A., Talior, I., Patel, N., Huang, X., Furtado, L.M., Bilan, P.J., Mann, M. & Klip, A. (2006) Insulin-dependent interactions of proteins with GLUT4 revealed through stable isotope labeling by amino acids in cell culture (SILAC). *J. Proteome Res.* **5**: 64–75.

Garavelli, J.S. (2004) The RESID Database of Protein Modifications as a resource and annotation tool. *Proteomics* **4**: 1527–1533.

Gerber, S.A., Kettenbach, A.N., Rush, J. & Gygi, S.P. (2007) The absolute quantification strategy: application to phosphorylation profiling of human separase serine 1126. *Methods Mol. Biol.* **359**: 71–86.

Gerber, S.A., Rush, J., Stemman, O., Kirschner, M.W. & Gygi, S.P. (2003) Absolute quantification of proteins and phosphoproteins from cell lysates by tandem MS. *Proc. Natl Acad. Sci. USA* **100**: 6940–6945.

Gharahdaghi, F., Weinberg, C.R., Meagher, D.A., Imai, B.S. & Mische, S.M. (1999) Mass spectrometric identification of proteins from silver-stained polyacrylamide gel: a method for the removal of silver ions to enhance sensitivity. *Electrophoresis* **20**: 601–605.

Gharbi, S., Gaffney, P., Yang, A., Zvelebil, M.J., Cramer, R., Waterfield, M.D. & Timms, J.F. (2002) Evaluation of two-dimensional differential gel electrophoresis for proteomic expression analysis of a model breast cancer cell system. *Mol. Cell Proteomics* **1**: 91–98.

Griffiths, J.R., Unwin, R.D., Evans, C.A., Leech, S.H., Corfe, B.M. & Whetton, A.D. (2007) The application of a hypothesis-driven strategy to the sensitive detection and location of acetylated lysine residues. *J. Am. Soc. Mass Spectrom.* **18**: 1423–1428.

Gruhler, A., Schulze, W.X., Matthiesen, R., Mann, M. & Jensen, O.N. (2005) Stable isotope labeling of Arabidopsis thaliana cells and quantitative proteomics by mass spectrometry. *Mol. Cell Proteomics* **4**: 1697–1709.

Guerrera, I.C., Predic-Atkinson, J., Kleiner, O., Soskic, V. & Godovac-Zimmermann, J. (2005) Enrichment of phosphoproteins for proteomic analysis using immobilized Fe(III)-affinity adsorption chromatography. *J. Proteome Res.* **4**: 1545–1553.

Gygi, S.P., Rist, B., Gerber, S.A., Turecek, F., Gelb, M.H. & Aebersold, R. (1999) Quantitative analysis of complex protein mixtures using isotope-coded affinity tags. *Nat. Biotechnol.* **17**: 994–999.

Hanahan, D. & Weinberg, R.A. (2000) The hallmarks of cancer. *Cell* **100**: 57–70.

Hart, S.R., Waterfield, M.D., Burlingame, A.L. & Cramer, R. (2002) Factors governing the solubilization of phosphopeptides retained on ferric NTA IMAC beads and their analysis by MALDI TOFMS. *J. Am. Soc. Mass Spectrom.* **13**: 1042–1051.

Harvey, D.J. (2005) Proteomic analysis of glycosylation: structural determination of N- and O-linked glycans by mass spectrometry. *Expert Rev. Proteomics* **2**: 87–101.

Haydon, C.E., Eyers, P.A., Aveline-Wolf, L.D., Resing, K.A., Maller, J.L. & Ahn, N.G. (2003) Identification of novel phosphorylation sites on Xenopus laevis Aurora A and analysis of phosphopeptide enrichment by immobilized metal-affinity chromatography. *Mol. Cell Proteomics* **2**: 1055–1067.

Heller, M., Ye, M., Michel, P.E., Morier, P., Stalder, D., Junger, M.A., Aebersold, R., Reymond, F. & Rossier, J.S. (2005) Added value for tandem mass spectrometry shotgun proteomics data validation through isoelectric focusing of peptides. *J. Proteome Res.* **4**: 2273–2282.

Henzel, W.J., Billeci, T.M., Stults, J.T., Wong, S.C., Grimley, C. & Watanabe, C. (1993) Identifying proteins from two-dimensional gels by molecular mass searching of peptide fragments in protein sequence databases. *Proc. Natl Acad. Sci. USA* **90**: 5011–5015.

Henzel, W.J., Watanabe, C. & Stults, J.T. (2003) Protein identification: the origins of peptide mass fingerprinting. *J. Am. Soc. Mass Spectrom.* **14**: 931–942.

Hermjakob, H. (2006) The HUPO Proteomics Standards Initiative - Overcoming the Fragmentation of Proteomics Data. *Proteomics* **6 Suppl. 2**: 34–38.

Hunter, T. (1998) The Croonian Lecture 1997. The phosphorylation of proteins on tyrosine: its role in cell growth and disease. *Phil. Trans. R. Soc. Lond. B. Biol. Sci.* **353**: 583–605.

Jones, P., Cote, R.G., Martens, L., Quinn, A.F., Taylor, C.F., Derache, W., Hermjakob, H. & Apweiler, R. (2006) PRIDE: a public repository of protein and peptide identifications for the proteomics community. *Nucleic Acids Res.* **34**: D659–663.

Kaji, H., Saito, H., Yamauchi, Y., Shinkawa, T., Taoka, M., Hirabayashi, J., Kasai, K., Takahashi, N. & Isobe, T. (2003) Lectin affinity capture, isotope-coded tagging and mass spectrometry to identify N-linked glycoproteins. *Nat. Biotechnol.* **21**: 667–672.

Keller, A., Nesvizhskii, A.I., Kolker, E. & Aebersold, R. (2002) Empirical statistical model to estimate the accuracy of peptide identifications made by MS/MS and database search. *Anal. Chem.* **74**: 5383–5392.

Keller, A., Eng, J., Zhang, N., Li, X.J. & Aebersold, R. (2005) A uniform proteomics MS/MS analysis platform utilizing open XML file formats. *Mol. Syst. Biol.* **1**: 2005 0017.

Kernec, F., Unlu, M., Labeikovsky, W., Minden, J.S. & Koretsky, A.P. (2001) Changes in the mitochondrial proteome from mouse hearts deficient in creatine kinase. *Physiol. Genomics* **6**: 117–128.

Kirkpatrick, D.S., Denison, C. & Gygi, S.P. (2005a) Weighing in on ubiquitin: the expanding role of mass-spectrometry-based proteomics. *Nat. Cell Biol.* **7**: 750–757.

Kirkpatrick, D.S., Gerber, S.A. & Gygi, S.P. (2005b) The absolute quantification strategy: a general procedure for the quantification of proteins and post-translational modifications. *Methods* **35**: 265–273.

Kito, K., Ota, K., Fujita, T. & Ito, T. (2007) A synthetic protein approach toward accurate mass spectrometric quantification of component stoichiometry of multi-protein complexes. *J. Proteome Res.* **6**: 792–800.

Kochetov, A.V. (2006) [Alternative translation start sites and their significance for eukaryotic proteome]. *Mol. Biol. (Mosk)* **40**: 788–795.

Krijgsveld, J., Ketting, R.F., Mahmoudi, T., Johansen, J., Artal-Sanz, M., Verrijzer, C.P., Plasterk, R.H. & Heck, A.J. (2003) Metabolic labeling of C. elegans and D. melanogaster for quantitative proteomics. *Nat. Biotechnol.* **21**: 927–931.

Krokhin, O., Ens, W., Standing, K.G., Wilkins, J. & Perreault, H. (2004) Site-specific N-glycosylation analysis: matrix-assisted laser desorption/ionization quadrupole-quadrupole time-of-flight tandem mass spectral signatures for recognition and identification of glycopeptides. *Rapid Commun. Mass Spectrom.* **18**: 2020–2030.

Larsen, M.R., Thingholm, T.E., Jensen, O.N., Roepstorff, P. & Jorgensen, T.J. (2005) Highly selective enrichment of phosphorylated peptides from peptide mixtures using titanium dioxide microcolumns. *Mol. Cell Proteomics* **4**: 873–886.

Le Blanc, J.C., Hager, J.W., Ilisiu, A.M., Hunter, C., Zhong, F. & Chu, I. (2003) Unique scanning capabilities of a new hybrid linear ion trap mass spectrometer (Q TRAP) used for high sensitivity proteomics applications. *Proteomics* **3**: 859–869.

Lee, C. & Roy, M. (2004) Analysis of alternative splicing with microarrays: successes and challenges. *Genome Biol.* **5**: 231.

Liao, Z., Cao, J., Li, S. *et al.* (2007) Proteomic and peptidomic analysis of the venom from Chinese tarantula Chilobrachys jingzhao. *Proteomics* **7**: 1892–1907.

Lilley, K.S. & Friedman, D.B. (2004) All about DIGE: quantification technology for differential-display 2D-gel proteomics. *Expert Rev. Proteomics* **1**: 401–409.

Lopez, J.L. (2007) Two-dimensional electrophoresis in proteome expression analysis. *J. Chromatogr. B. Analyt. Technol. Biomed. Life Sci.* **849**: 190–202.

Lopez, M.F., Berggren, K., Chernokalskaya, E., Lazarev, A., Robinson, M. & Patton, W.F. (2000) A comparison of silver stain and SYPRO Ruby Protein Gel Stain with respect to protein detection in two-dimensional gels and identification by peptide mass profiling. *Electrophoresis* **21**: 3673–3683.

Makrantoni, V., Antrobus, R., Botting, C.H. & Coote, P.J. (2005) Rapid enrichment and analysis of yeast phosphoproteins using affinity chromatography, 2D-PAGE and peptide mass fingerprinting. *Yeast* **22**: 401–414.

Mann, M. & Jensen, O.N. (2003) Proteomic analysis of post-translational modifications. *Nat. Biotechnol.* **21**: 255–261.

Mann, M., Ong, S.E., Gronborg, M., Steen, H., Jensen, O.N. & Pandey, A. (2002) Analysis of protein phosphorylation using mass spectrometry: deciphering the phosphoproteome. *Trends Biotechnol.* **20**: 261–268.

Mayya, V., Rezual, K., Wu, L., Fong, M.B. & Han, D.K. (2006) Absolute quantification of multisite phosphorylation by selective reaction monitoring mass spectrometry: determination of inhibitory phosphorylation status of cyclin-dependent kinases. *Mol. Cell Proteomics* **5**: 1146–1157.

Mazanek, M., Mituloviae, G., Herzog, F., Stingl, C., Hutchins, J.R., Peters, J.M. & Mechtler, K. (2007) Titanium dioxide as a chemo-affinity solid phase in offline phosphopeptide chromatography prior to HPLC-MS/MS analysis. *Nat. Protoc.* **2**: 1059–1069.

Mikesh, L.M., Ueberheide, B., Chi, A., Coon, J.J., Syka, J.E., Shabanowitz, J. & Hunt, D.F. (2006) The utility of ETD mass spectrometry in proteomic analysis. *Biochim. Biophys. Acta* **1764**: 1811–1822.

Nesvizhskii, A.I., Keller, A., Kolker, E. & Aebersold, R. (2003) A statistical model for identifying proteins by tandem mass spectrometry. *Anal. Chem.* **75:** 4646–4658.

Neville, D.C., Rozanas, C.R., Price, E.M., Gruis, D.B., Verkman, A.S. & Townsend, R.R. (1997) Evidence for phosphorylation of serine 753 in CFTR using a novel metal-ion affinity resin and matrix-assisted laser desorption mass spectrometry. *Protein Sci.* **6:** 2436–2445.

Nicholas, B., Skipp, P., Mould, R., Rennard, S., Davies, D.E., O'Connor, C.D. & Djukanovic, R. (2006) Shotgun proteomic analysis of human-induced sputum. *Proteomics* **6:** 4390–4401.

Nielsen, M.L., Savitski, M.M. & Zubarev, R.A. (2006) Extent of modifications in human proteome samples and their effect on dynamic range of analysis in shotgun proteomics. *Mol. Cell Proteomics* **5:** 2384–2391.

Nuhse, T.S., Stensballe, A., Jensen, O.N. & Peck, S.C. (2003) Large-scale analysis of in vivo phosphorylated membrane proteins by immobilized metal ion affinity chromatography and mass spectrometry. *Mol. Cell Proteomics* **2:** 1234–1243.

Olsen, J.V., Andersen, J.R., Nielsen, P.A., Nielsen, M.L., Figeys, D., Mann, M. & Wisniewski, J.R. (2004) HysTag—a novel proteomic quantification tool applied to differential display analysis of membrane proteins from distinct areas of mouse brain. *Mol. Cell Proteomics* **3:** 82–92.

Olsen, J.V., Blagoev, B., Gnad, F., Macek, B., Kumar, C., Mortensen, P. & Mann, M. (2006) Global, in vivo, and site-specific phosphorylation dynamics in signaling networks. *Cell* **127:** 635–648.

Ong, S.E. & Mann, M. (2006) A practical recipe for stable isotope labeling by amino acids in cell culture (SILAC). *Nat. Protoc.* **1:** 2650–2660.

Ong, S.E. & Mann, M. (2007) Stable isotope labeling by amino acids in cell culture for quantitative proteomics. *Methods Mol. Biol.* **359:** 37–52.

Ong, S.E., Foster, L.J. & Mann, M. (2003) Mass spectrometric-based approaches in quantitative proteomics. *Methods* **29:** 124–130.

Ong, S.E., Blagoev, B., Kratchmarova, I., Kristensen, D.B., Steen, H., Pandey, A. & Mann, M. (2002) Stable isotope labeling by amino acids in cell culture, SILAC, as a simple and accurate approach to expression proteomics. *Mol. Cell Proteomics* **1:** 376–386.

Orchard, S., Jones, P., Taylor, C., Zhu, W., Julian, R.K., Jr., Hermjakob, H. & Apweiler, R. (2007) Proteomic data exchange and storage: the need for common standards and public repositories. *Methods Mol. Biol.* **367:** 261–270.

Oyama, M., Kozuka-Hata, H., Suzuki, Y., Semba, K., Yamamoto, T. & Sugano, S. (2007) Diversity of translation start sites may define increased complexity of the human short ORFeome. *Mol. Cell Proteomics* **6:** 1000–1006.

Paizs, B. & Suhai, S. (2005) Fragmentation pathways of protonated peptides. *Mass Spectrom. Rev.* **24:** 508–548.

Papageorgopoulos, C., Caldwell, K., Shackleton, C., Schweingrubber, H. & Hellerstein, M.K. (1999) Measuring protein synthesis by mass isotopomer distribution analysis (MIDA). *Anal. Biochem.* **267:** 1–16.

Papageorgopoulos, C., Caldwell, K., Schweingrubber, H., Neese, R.A., Shackleton, C.H. & Hellerstein, M. (2002) Measuring synthesis rates of muscle creatine kinase and myosin with stable isotopes and mass spectrometry. *Anal. Biochem.* **309:** 1–10.

Pappin, D.J., Hojrup, P. & Bleasby, A.J. (1993) Rapid identification of proteins by peptide-mass fingerprinting. *Curr. Biol.* **3**: 327–332.

Pedrioli, P.G., Eng, J.K., Hubley, R. *et al.* (2004) A common open representation of mass spectrometry data and its application to proteomics research. *Nat. Biotechnol.* **22**: 1459–1466.

Peng, J., Elias, J.E., Thoreen, C.C., Licklider, L.J. & Gygi, S.P. (2003a) Evaluation of multidimensional chromatography coupled with tandem mass spectrometry (LC/LC-MS/MS) for large-scale protein analysis: the yeast proteome. *J. Proteome Res.* **2**: 43–50.

Peng, J., Schwartz, D., Elias, J.E., Thoreen, C.C., Cheng, D., Marsischky, G., Roelofs, J., Finley, D. & Gygi, S.P. (2003b) A proteomics approach to understanding protein ubiquitination. *Nat. Biotechnol.* **21**: 921–926.

Perkins, D.N., Pappin, D.J., Creasy, D.M. & Cottrell, J.S. (1999) Probability-based protein identification by searching sequence databases using mass spectrometry data. *Electrophoresis* **20**: 3551–3567.

Pinkse, M.W., Uitto, P.M., Hilhorst, M.J., Ooms, B. & Heck, A.J. (2004) Selective isolation at the femtomole level of phosphopeptides from proteolytic digests using 2D-NanoLC-ESI-MS/MS and titanium oxide precolumns. *Anal. Chem.* **76**: 3935–3943.

Pratt, J.M., Robertson, D.H., Gaskell, S.J. *et al.* (2002) Stable isotope labelling in vivo as an aid to protein identification in peptide mass fingerprinting. *Proteomics* **2**: 157–163.

Pratt, J.M., Simpson, D.M., Doherty, M.K., Rivers, J., Gaskell, S.J. & Beynon, R.J. (2006) Multiplexed absolute quantification for proteomics using concatenated signature peptides encoded by QconCAT genes. *Nat. Protoc.* **1**: 1029–1043.

Reynolds, K.J., Yao, X. & Fenselau, C. (2002) Proteolytic 18O labeling for comparative proteomics: evaluation of endoprotease Glu-C as the catalytic agent. *J. Proteome Res.* **1**: 27–33.

Rivers, J., Simpson, D.M., Robertson, D.H., Gaskell, S.J. & Beynon, R.J. (2007) Absolute multiplexed quantitative analysis of protein expression during muscle development using QconCAT. *Mol. Cell Proteomics* **6**: 1416–1427.

Roepstorff, P. & Fohlman, J. (1984) Proposal for a Common Nomenclature for Sequence Ions in Mass-Spectra of Peptides. *Biomed. Mass Spectrom.* **11**: 601.

Salomon, A.R., Ficarro, S.B., Brill, L.M. *et al.* (2003) Profiling of tyrosine phosphorylation pathways in human cells using mass spectrometry. *Proc. Natl Acad. Sci. USA* **100**: 443–448.

Schagger, H., Aquila, H. & Von Jagow, G. (1988) Coomassie blue-sodium dodecyl sulfate-polyacrylamide gel electrophoresis for direct visualization of polypeptides during electrophoresis. *Anal. Biochem.* **173**: 201–205.

Schmidt, A., Kellermann, J. & Lottspeich, F. (2005) A novel strategy for quantitative proteomics using isotope-coded protein labels. *Proteomics* **5**: 4–15.

Schroeder, M.J., Webb, D.J., Shabanowitz, J., Horwitz, A.F. & Hunt, D.F. (2005) Methods for the detection of paxillin post-translational modifications and interacting proteins by mass spectrometry. *J. Proteome Res.* **4**: 1832–1841.

Schwartz, B.L. & Bursey, M.M. (1992) Some proline substituent effects in the tandem mass-spectrum of protonated pentaalanine. *Biol. Mass Spectrom.* **21**: 92–96.

Shadforth, I. & Bessant, C. (2006) Genome annotating proteomics pipelines: available tools. *Expert Rev. Proteomics* **3**: 621–629.

Shadforth, I., Xu, W., Crowther, D. & Bessant, C. (2006) GAPP: a fully automated software for the confident identification of human peptides from tandem mass spectra. *J. Proteome Res.* **5**: 2849–2852.

Shaw, J., Rowlinson, R., Nickson, J., Stone, T., Sweet, A., Williams, K. & Tonge, R. (2003) Evaluation of saturation labelling two-dimensional difference gel electrophoresis fluorescent dyes. *Proteomics* **3**: 1181–1195.

Shevchenko, A., Wilm, M., Vorm, O. & Mann, M. (1996) Mass spectrometric sequencing of proteins silver-stained polyacrylamide gels. *Anal. Chem.* **68**: 850–858.

Shevchenko, A., Sunyaev, S., Loboda, A., Bork, P., Ens, W. & Standing, K.G. (2001) Charting the proteomes of organisms with unsequenced genomes by MALDI-quadrupole time-of-flight mass spectrometry and BLAST homology searching. *Anal. Chem.* **73**: 1917–1926.

Silva, J.C., Gorenstein, M.V., Li, G.Z., Vissers, J.P. & Geromanos, S.J. (2006) Absolute quantification of proteins by LCMSE: a virtue of parallel MS acquisition. *Mol. Cell Proteomics* **5**: 144–156.

Skotheim, R.I. & Nees, M. (2007) Alternative splicing in cancer: Noise, functional, or systematic? *Int. J. Biochem. Cell Biol.* **39**: 1432–1449.

Smith, R.D., Anderson, G.A., Lipton, M.S., Pasa-Tolic, L., Shen, Y., Conrads, T.P., Veenstra, T.D. & Udseth, H.R. (2002) An accurate mass tag strategy for quantitative and high-throughput proteome measurements. *Proteomics* **2**: 513–523.

Steen, H., Kuster, B., Fernandez, M., Pandey, A. & Mann, M. (2001) Detection of tyrosine phosphorylated peptides by precursor ion scanning quadrupole TOF mass spectrometry in positive ion mode. *Anal. Chem.* **73**: 1440–1448.

Stensballe, A. & Jensen, O.N. (2004) Phosphoric acid enhances the performance of Fe(III) affinity chromatography and matrix-assisted laser desorption/ionization tandem mass spectrometry for recovery, detection and sequencing of phosphopeptides. *Rapid Commun. Mass Spectrom.* **18**: 1721–1730.

Stensballe, A., Jensen, O.N., Olsen, J.V., Haselmann, K.F. & Zubarev, R.A. (2000) Electron capture dissociation of singly and multiply phosphorylated peptides. *Rapid Commun. Mass Spectrom.* **14**: 1793–1800.

Strittmatter, E.F., Ferguson, P.L., Tang, K. & Smith, R.D. (2003) Proteome analyses using accurate mass and elution time peptide tags with capillary LC time-of-flight mass spectrometry. *J. Am. Soc. Mass Spectrom.* **14**: 980–991.

Swatton, J.E., Prabakaran, S., Karp, N.A., Lilley, K.S. & Bahn, S. (2004) Protein profiling of human postmortem brain using 2-dimensional fluorescence difference gel electrophoresis (2-D DIGE). *Mol. Psych.* **9**: 128–143.

Syka, J.E., Coon, J.J., Schroeder, M.J., Shabanowitz, J. & Hunt, D.F. (2004) Peptide and protein sequence analysis by electron transfer dissociation mass spectrometry. *Proc. Natl Acad. Sci. USA* **101**: 9528–9533.

Taylor, C.F., Paton, N.W., Garwood, K.L. *et al.* (2003) A systematic approach to modeling, capturing, and disseminating proteomics experimental data. *Nat. Biotechnol.* **21**: 247–254.

Thiede, B., Kretschmer, A. & Rudel, T. (2006) Quantitative proteome analysis of CD95 (Fas/Apo-1)-induced apoptosis by stable isotope labeling with amino acids in cell culture, 2-DE and MALDI-MS. *Proteomics* **6**: 614–622.

Thingholm, T.E., Jorgensen, T.J., Jensen, O.N. & Larsen, M.R. (2006) Highly selective enrichment of phosphorylated peptides using titanium dioxide. *Nat. Protoc.* 1: 1929–1935.

Thompson, A.J., Hart, S.R., Franz, C., Barnouin, K., Ridley, A. & Cramer, R. (2003) Characterization of protein phosphorylation by mass spectrometry using immobilized metal ion affinity chromatography with on-resin beta-elimination and Michael addition. *Anal. Chem.* 75: 3232–3243.

Unlu, M., Morgan, M.E. & Minden, J.S. (1997) Difference gel electrophoresis: a single gel method for detecting changes in protein extracts. *Electrophoresis* 18: 2071–2077.

Unwin, R.D., Griffiths, J.R., Leverentz, M.K., Grallert, A., Hagan, I.M. & Whetton, A.D. (2005) Multiple reaction monitoring to identify sites of protein phosphorylation with high sensitivity. *Mol. Cell Proteomics* 4: 1134–1144.

Unwin, R.D., Smith, D.L., Blinco, D. *et al.* (2006) Quantitative proteomics reveals posttranslational control as a regulatory factor in primary hematopoietic stem cells. *Blood* 107: 4687–4694.

Vaisar, T. & Urban, J. (1996) Probing the proline effect in CID of protonated peptides. *J. Mass Spectrom.* 31: 1185–1187.

Viswanathan, S., Unlu, M. & Minden, J.S. (2006) Two-dimensional difference gel electrophoresis. *Nat. Protoc.* 1: 1351–1358.

Vogt, J.A., Hunzinger, C., Schroer, K. *et al.* (2005) Determination of fractional synthesis rates of mouse hepatic proteins via metabolic 13C-labeling, MALDI-TOF MS and analysis of relative isotopologue abundances using average masses. *Anal. Chem.* 77: 2034–2042.

Vosseller, K., Trinidad, J.C., Chalkley, R.J. *et al.* (2006) O-linked N-acetylglucosamine proteomics of postsynaptic density preparations using lectin weak affinity chromatography and mass spectrometry. *Mol. Cell Proteomics* 5: 923–934.

Waridel, P., Frank, A., Thomas, H., Surendranath, V., Sunyaev, S., Pevzner, P. & Shevchenko, A. (2007) Sequence similarity-driven proteomics in organisms with unknown genomes by LC-MS/MS and automated de novo sequencing. *Proteomics* 7: 2318–2329.

Warwood, S., Mohammed, S., Cristea, I.M., Evans, C., Whetton, A.D. & Gaskell, S.J. (2006) Guanidination chemistry for qualitative and quantitative proteomics. *Rapid Commun. Mass Spectrom.* 20: 3245–3256.

Washburn, M.P., Wolters, D. & Yates, J.R., 3rd (2001) Large-scale analysis of the yeast proteome by multidimensional protein identification technology. *Nat. Biotechnol.* 19: 242–247.

Wiener, M.C., Sachs, J.R., Deyanova, E.G. & Yates, N.A. (2004) Differential mass spectrometry: a label-free LC-MS method for finding significant differences in complex peptide and protein mixtures. *Anal. Chem.* 76: 6085–6096.

Wiese, S., Reidegeld, K.A., Meyer, H.E. & Warscheid, B. (2007) Protein labeling by iTRAQ: a new tool for quantitative mass spectrometry in proteome research. *Proteomics* 7: 340–350.

Wilkins, M.R., Appel, R.D., Van Eyk, J.E. *et al.* (2006) Guidelines for the next 10 years of proteomics. *Proteomics* 6: 4–8.

Xie, H., Bandhakavi, S., Roe, M.R. & Griffin, T.J. (2007) Preparative peptide isoelectric focusing as a tool for improving the identification of lysine-acetylated peptides from complex mixtures. *J. Proteome Res.* 6: 2019–2026.

Yu, W., Vath, J.E., Huberty, M.C. & Martin, S.A. (1993) Identification of the facile gas-phase cleavage of the Asp Pro and Asp Xxx peptide-bonds in matrix-assisted laser-desorption time-of-flight mass-spectrometry. *Anal. Chem.* **65:** 3015–3023.

Yu, Y., Hoffhines, A.J., Moore, K.L. & Leary, J.A. (2007) Determination of the sites of tyrosine O-sulfation in peptides and proteins. *Nat. Methods* **4:** 583–588.

Zabrouskov, V., Senko, M.W., Du, Y., Leduc, R.D. & Kelleher, N.L. (2005) New and automated MSn approaches for top-down identification of modified proteins. *J. Am. Soc. Mass Spectrom.* **16:** 2027–2038.

Zhang, G., Spellman, D.S., Skolnik, E.Y. & Neubert, T.A. (2006) Quantitative phosphotyrosine proteomics of EphB2 signaling by stable isotope labeling with amino acids in cell culture (SILAC). *J. Proteome Res.* **5:** 581–588.

Zhou, H., Watts, J.D. & Aebersold, R. (2001) A systematic approach to the analysis of protein phosphorylation. *Nat. Biotechnol.* **19:** 375–378.

Zubarev, R.A., Kelleher, N.L. & McLafferty, F.W. (1998) Electron capture dissociation for structural characterization of multiply charged protein cations. *J. Am. Chem. Soc.* **120:** 3265–3266.

Zubarev, R.A., Horn, D.M., Fridriksson, E.K., Kelleher, N.L., Kruger, N.A., Lewis, M.A., Carpenter, B.K. & McLafferty, F.W. (2000) Electron capture dissociation of multiply charged protein cations. A nonergodic process. *Anal. Chem.* **72:** 563–573.

4

Vertical systems biology: From DNA to flux and back

Annamaria Bevilacqua, Stephen J. Wilkinson,
Richard Dimelow, Ettore Murabito, Samrina Rehman,
Maria Nardelli, Karen van Eunen, Sergio Rossell,
Frank J. Bruggeman, Nils Blüthgen, Dirk De Vos,
Jildau Bouwman, Barbara M. Bakker and
Hans V. Westerhoff.

Systems biology has witnessed exponential growth since its inception early this century, when genomics combined with mathematical biology. Much of systems biology has remained tightly linked with a single genomics methodology and with the observation of patterns at a single level of cell function, that is, the genome, the transcriptome, the proteome or the metabolome. In this chapter we shall describe a novel methodology that is orthogonal to these approaches. It analyses practically how functions in living organisms are being regulated at the same time at all these various levels of the cellular regulation hierarchy. The principles behind the methodology are reviewed and extended. This underpins complete time-dependent hierarchical regulation analysis from flux through metabolites, enzyme, mRNA, gene and back.

An important part of the methodology, discriminating between metabolic and gene-expression regulation, is then illustrated in detail experimentally. It is shown how yeast regulates its capacity to produce ethanol when it is confronted with a lack of nitrogen source. It regulates the flux through alcohol dehydrogenase first by reducing the gene expression of the corresponding enzyme. As time proceeds, the close to 100% gene-expression regulation, is replaced with virtually entirely metabolic regulation.

How a gene-expression regulatory coefficient may come about is then illustrated in an *in silico* model for yeast glycolysis. The expression level of the glucose transporter is supposed to be regulated at the level of protein synthesis by the pyruvate concentration. This leads to regulation of glucose uptake partly through gene-expression, partly through direct metabolic interactions.

1 Hierarchies in control: Living the dogma of molecular biology

More than 60 years of molecular biology and biochemistry have led to a clear view: information flows from a library in which much of it remains stably encoded, through labile messenger molecules, through often stable molecular machines, to processes that are often functional for the living cell. *Figure 1* illustrates this for single cells with

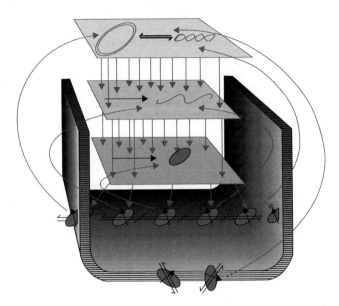

Figure 1. *The hierarchical subdivision of cellular biochemistry. From top to bottom, the levels are those of DNA, mRNA, proteins, and metabolism. From www.siliconcell.net, with permission.*

an emphasis on metabolic processes as 'functions'. The downward arrows refer to the direction of influence that has become associated with this paradigm: DNA determines mRNA determines proteins which determine function.

From the early days it was realized by some that this was not always fully the case: in the case of the *lac* operon the expression of the information inscribed on the DNA depended on the concentration of allolactose in *E. coli*. The upward, reverse arrows in *Figure 1* reflect this additional 'reverse' paradigm of regulation of gene expression.

The biochemical activities of various enzymes in metabolic pathways are regulated in multifarious ways by the concentrations of metabolites. This suggests that such metabolic regulation might also be used in actual cases of regulation *in vivo*. How would it otherwise have withstood evolutionary pressure? For a long time there was no conceptual framework in which regulation or even control at multiple points in a network could be appreciated, other than in qualitative and unbalanced ways.

This was true initially for the control of fluxes through metabolic pathways. Biochemistry had been quite successful in identifying, isolating and then characterizing the molecular machines (enzymes) that carried out the chemical conversions that were necessary for biological function. This included those that made it possible that the Gibbs energy present in catabolic substrates be delivered in neat quanta corresponding to the Gibbs free energy of hydrolysis of ATP or the proton motive force (Westerhoff and Van Dam, 1987). However, biochemistry itself was not quite ready to deal with the fact that any function carried out *in vivo* would depend on a number of such molecular machines at the same time. ATP production by glycolysis would not only depend on the molecular machines directly phosphorylating ADP (i.e. pyruvate kinase and phosphoglycerate kinase) but also on the hexokinase and aldolase supplying the former two with carbon substrate. It was really thanks to scientists such as Heinrich (Heinrich and Rapoport, 1974; Heinrich *et al.*, 1977), Kacser (Kacser and Burns, 1973), and Groen (Groen *et al.*,1982) that a methodology was developed and applied experimentally that enabled the appreciation that the flux through an important pathway was controlled by many of the enzymes in that pathway, notably to different extents. This Metabolic Control Analysis was an early form of practical systems biology, as its definitions were operational (Groen et *al.*, 1982; Middleton and Kacser, 1983) and it showed that the control exerted by one enzyme on the flux also depended on the activities of the other enzymes in the pathway (Wanders *et al.*, 1984).

Metabolic Control Analysis (MCA) was almost entirely developed for, and by, biochemists who were exclusively interested in metabolic regulation. It took a relatively long time, therefore, for an extension to be developed that also accounted for the additional complexity that metabolic processes might also be regulated at the level of the transcription of DNA into mRNA or at the level of translation of mRNA into protein. The first such extension of MCA came in 1990 (Westerhoff and Van Workum, 1990; Westerhoff *et al.*, 1990) with a further formalization and generalization by Kahn and Westerhoff (1991). The simplest case considered was that of dictatorial control, that is, the case where the upward arrows of *Figure 1* were absent. This perhaps corresponded most with the dogma of molecular biology. Yet it showed that control of metabolic flux by transcription did not exclude control of the same flux by a metabolic enzyme. We may illustrate this for the case of the two-enzyme pathway of *Figure 2*. We shall do this for the ultimate steady state, that is, the condition where the concentrations of mRNA, protein and metabolite X have become independent of time. Concentrations of substrate S, product P, ribonucleotide tri-phosphates, aminoacyl tRNAs, amino acids, ribonucleotides, ribosome and ATP are assumed to be fixed by the rest of the metabolism or the cell's environment.

An important aspect that systems biology brings to biology is the importance of being precise. How would one, for the scheme of *Figure 2*, discuss whether the metabolic flux is controlled by transcription or by metabolism? Knocking out the gene encoding protein 2 would eliminate the flux through the metabolic pathway, unless there was a parallel pathway leading to the same product P, or a second gene encoding an enzyme with similar specificity. Eliminating enzyme 1 or enzyme 2 would similarly eliminate the production flux of pathway P. Eliminating the proteolytic breakdown of enzyme 2 would increase the flux, but to an unclear extent as the concentration of intermediary metabolite X might soon become limiting. One way out would be to declare that all processes in *Figure 2* would limit or determine the production flux of P, but as a pathway such as that in *Figure 2* will always be embedded in the rest of cell metabolism, this would mean that for yeast, the flux would be controlled by 6000 genes, 6000 mRNA degradations and 6000 proteins at the same time, and that all of these would be equally important.

Systems biology focuses on aspects where systems of the biological macromolecules differ functionally from the individual macromolecules (Alberghina and Westerhoff, 2005; Boogerd *et al.*, 2007). Molecular biology rather looks at the molecular function of each of the macromolecules independently. The flux through an enzyme in isolation is (usually; exceptions can be accommodated, see below) proportional to the concentration of that enzyme. Molecular biology, being as successful as

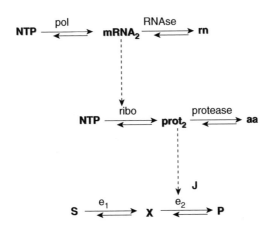

Figure 2. *A diagram illustrating the regulation of a metabolic flux, J, through a two-step pathway converting substrate S, via intermediate X to P at a steady state rate J. The second step in the pathway is catalysed by protein 2 which is identical to the enzyme 2 catalysing this step. Protein 2 is synthesized by ribosomes and degraded by the proteases in the cell (or diluted as a result of population growth) where the former process is instructed by the corresponding mRNA₂. This mRNA is synthesized by RNA polymerase and degraded by RNases. This picture corresponds to dictatorial control, as there are no upward arrows, which would reflect control of transcription or translation by metabolite X.*

it is for understanding the molecular mechanisms of living organisms, has led many to assume that its concepts are germane for networks. However, from the molecular biology perspective alone the understanding of the control of the two steady-state fluxes through the metabolic pathway in *Figure 2* is paradoxical. Focusing on the steady-state flux through enzyme 1 one might stipulate that this flux should be proportional to the concentration of enzyme 1. And, focusing on the flux through enzyme 2 one might propose that that flux should be proportional to the concentration of enzyme 2. However, at steady states, the fluxes should always be equal to each other. Therefore, if one doubles the concentration of enzyme 1 and keeps the concentration of enzyme 2 in *Figure 2* constant, the view that molecular biology determines everything would seem to predict the same steady-state flux to double and to stay constant at the same time.

It is not that molecular biology is wrong of course, but that it was not made to deal with systems. For systems it is incomplete: a view of one molecule at a time is not able to predict the flux through a metabolic pathway. The resolution of the above paradox is that if one doubles the concentration of enzyme 2, then the environment of enzyme 2 will also change and the latter change will also affect enzyme 2. In this example the change is a decrease in the concentration of X, which will reduce the rate at which enzyme 2 functions after it has doubled. The same decrease in [X] will lead to a smaller product inhibition effect on enzyme 1 hence increase the rate through that enzyme. The decrease in the doubled enzyme rate 2 and the increase in the enzyme rate 1 will both continue until the two rates become equal again. Then the effect of the doubling of the concentration of enzyme 2 will have become the same on rate v_1 and rate v_2, but the effect will be less than a doubling of the flux and more than constancy. The precise effect on the flux is a system property. It has been defined (Burns *et al.*, 1985) as the flux control coefficient, which is virtually equal to the percentage increase in the flux for a 1% increase in the concentration of enzyme 2 (Westerhoff and Van Dam, 1987):

$$C_{e_2}^{J} = \left(\frac{d \ln J}{d \ln e_2} \right)_{e_1, other\ parameters, system\ steady\ state} = \left(\frac{dJ\!\!\Big/\!J}{de_2\!\!\Big/\!e_2} \right)_{e_1, other\ parameters, system\ steady\ state} = \tag{1}$$

$$\cong \% \textit{increase in flux J if } 1\% \textit{increase in} [e_2]$$

It turns out that total flux control must equal 1, that is, the sum of the control exerted by enzyme 1 and the control exerted by enzyme 2 on flux *J* must add up to 1 (Heinrich and Rapoport, 1974; Kacser and Burns, 1973).

As may be expected from the above account of what led to the new steady state after doubling of the concentration of enzyme 2, the precise distribution of flux control between the two enzymes depends on the differential extent to which the two enzymes respond to changes in [X]. The same applies to the control the two enzymes exert on the steady state concentration of X (Kacser and Burns, 1973; Westerhoff and Kell, 1987).

MCA tells us that the flux control coefficients of enzymes 1 and 2 of *Figure 2* must sum to unity. The issue not addressed by MCA, but of particular relevance to this chapter, is whether the fact that the total control of 1 has now already been attributed to enzyme 1 and 2 implies that there is no control possible by RNA polymerase, RNase, the ribosomes or protease. This would seem odd as a general result. One should certainly expect that in some cases transcription should control important metabolic fluxes, or indeed the most important metabolic flux, that is, the flux of biomass synthesis, growth rate.

Hierarchical Control Analysis (HCA; Kahn and Westerhoff, 1991; Westerhoff and Van Workum, 1990; Westerhoff *et al.*, 1990) resolved this paradox. It clarified that transcription plus RNase together must exert a control of zero on the metabolic flux, but these two processes may well exert strong but opposite flux control individually. In the case where RNase is proportionally dependent on the concentration of $mRNA_2$ and transcription is independent of that concentration, transcription should control the metabolic flux at a flux control coefficient of 1. $mRNA_2$ degradation should then control the flux at a coefficient of minus 1. The –1 means that a 1% activation of $mRNA_2$ breakdown rate would reduce the metabolic flux by 1%. Similarly translation and proteolysis are likely to control the metabolic flux at flux control coefficients of 1 and –1, respectively.

In this way the dogma of molecular biology that DNA determines mRNA determines protein determines function is translated to a series of control coefficients at the levels of intermediary metabolism, protein metabolism and mRNA metabolism. Implied by the dogma itself, the simplest realistic case has control distributed over the various levels: Metabolic control of cell function is not in contradiction to transcription control. Also important is the implication that RNA degradation and protein degradation are equally important control points of cell function as are transcription and translation (Kahn and Westerhoff, 1991; Westerhoff *et al.*, 1990).

All these conclusions of distributed control were already reached for a situation without feedback from metabolism to transcription. This absence of feedback is reflected by the fact that in such a 'dictatorial' control hierarchy, metabolic processes do not exert control on transcription or translation. This becomes different when such feedbacks exist. In such democratic control hierarchies, all control coefficients become attenuated (or amplified, depending on the sign of the loop) by the same circular regulation ('reverberation') coefficient that consists of the multiplication of control and elasticity coefficients along the loop (Kahn and Westerhoff, 1991; Hofmeyr and Westerhoff, 2001; van der Gugten and Westerhoff, 1997; Westerhoff *et al.*, 1990). Typically, that circular attenuation will not be in place at short time scales; after a perturbation the system may first respond fairly strongly followed by a relaxation back, so as to end closer to the steady state before the perturbation.

Snoep *et al.* (2002) applied HCA experimentally to DNA supercoiling in *E. coli*. It turned out this was controlled mildly by DNA gyrase and by topoisomerase I. Part of the homeostasis was due to the direct feedback of supercoiling on DNA gyrase action and part was due to a regulatory loop in which gyrase and topoisomerase gene expression was affected by DNA supercoiling. Here decreased DNA supercoiling led to enhanced transcription of the DNA gyrase genes and an enhanced synthetic flux of DNA gyrase, which in turn enhanced DNA supercoiling.

The control discussed in these control analyses refers to the potential for change, independent of whether that potential is brought into effect: A control coefficient reflects the change in a variable if one were to modulate an enzyme activity or concentration in the system, independently of whether one actually does so. In fact many of the possible control points in the cell with substantial control coefficients may never be modulated, either by the experimentalist or by the cell itself.

2 Hierarchies in regulation: Hierarchical Regulation Analysis

One issue to be resolved by fundamental science is which parts of the living cell are most critical for its functioning. This issue is addressed by the control analyses described in the previous section. Genomics analyses do not yet address these issues. They tend to be limited to describing how the living organism responds to challenges to which it is exposed. Top-down systems biology tries to recognize patterns in that response, and infer the mechanisms of its regulation (Alberghina and Westerhoff, 2005). That inference is, however, based on phenomenological models such as ones that assume that cell function is regulated exclusively through genetic networks. Such genetic networks are defined in terms of nodes that refer to mRNAs and edges that refer to the latter influencing the synthesis of other such mRNAs. There is no reference to communication through a large metabolic network consisting of many nodes that are flux connected, hence subject to flux balancing conditions at steady state. The effect may well be that regulatory links are deduced from the analyses that have no physical meaning and that will appear to change with conditions even if in reality regulation does not change.

In an integration between top-down and bottom-up systems biology Palsson and colleagues have engaged in reconstruction of intracellular networks based on genome sequencing data and on what is known about biochemical pathways and the chemistry required to complete metabolic conversions (Palsson, 2006). In addition for each of the biochemical reaction steps genomics enables the identification of the corresponding protein and mRNA, following the 'dogma' of molecular biology (see above). Much of the networks of cell function are thus becoming identified. Consequently, it may seem futile for analyses of genomics data to remain blind to what is already known about the networks in the living organism.

The bottom-up systems biology approach (Rossell *et al.*, 2005, 2006; Ter Kuile and Westerhoff, 2001) that has been called Hierarchical Regulation Analysis (HRA) is not blind to what is already known about possible mechanisms of regulation. It enables the dissection of the regulation of a process *in vivo* into its regulatory components.

The first aim of this procedure was to determine whether the approach itself is at all necessary. For, if in all cases regulation was exclusively through the adjustment of transcription, or exclusively through alterations in metabolic interactions, then such a dissection would not be useful. In the above-quoted work, but also in Even *et al.* (2003) and Suarez *et al.* (2005), it turned out that regulation was neither exclusively transcriptional nor exclusively metabolic.

It is important to clarify the distinction between regulation analysis and MCA/HCA. In regulation analysis, the organism under study is regulating its own metabolism, usually as a response to an external change, but occasionally also with the progression of an internal clock. The latter case obtains during the cell cycle, but also in developmental biology. To adjust the flux through a process, the organism may increase the concentration of the substrate or an allosteric modifier of the enzyme catalysing that process. This corresponds to metabolic regulation of that process. The organism may also increase the level of the enzyme that catalyses the process, corresponding to gene-expression regulation of that process. Or, the organism may activate signal transduction leading to phosphorylation of the enzyme thereby increasing its V_{max}. The latter two cases combined have been called hierarchical regulation (Ter Kuile and Westerhoff, 2001). In all these cases the rate of the reaction catalysed by the enzyme will increase as a result of the regulatory activities of the organism.

HRA studies the regulation of the rate of a process in the living cell by that cell. HRA examines which of the factors that affect the rate of the process under study the cell actually activates. In contrast, HCA studies the extent to which activation of each enzyme in the network would have led to an increase in pathway flux, had that enzyme indeed been activated; HCA determines the extent to which the various enzymes in the network limit the pathway flux. When required to regulate the rate of a process (say process 2 in *Figure 2*), the cell may decide to activate the enzyme with the highest control coefficient (0.7, see above) by activating its gene expression by 10%. This would lead to an increase in flux through the pathway of 7%. The hierarchical regulation coefficient of this process (see below) would then equal 10/7 = 1.4. The substrate of this enzyme would decrease strongly and its product would increase, leading to a counteracting, homeostatic metabolic regulation of –3/7 = –0.4. The cell may not activate the other enzymes in the pathway, with the flux control coefficient of 0.3. The process catalysed by that enzyme would then have a hierarchical regulation coefficient of 0.0 and a metabolic regulation coefficient of 1.0.

Just as HCA was preceded by MCA, HRA had a predecessor that dealt exclusively with the dissection of metabolic regulation into its components (Sauro, 1990). We exemplify this again for the regulation of the rate of the process 1 in *Figure 2*.

We assume that enzyme 1 obeys the product inhibition form of the Michaelis–Menten equation:

$$v_1 = [e_1] \cdot k_{cat,1} \cdot \frac{\dfrac{[S]}{K_S}}{1 + \dfrac{[S]}{K_S} + \dfrac{[X]}{K_X}} \tag{2}$$

The dependence of the rate of the process 1 on the various metabolic factors regulating it is then given by:

$$\% \text{ change in } v_i = \% \text{ change in } e_i + \varepsilon_S \cdot \% \text{ change in } [S] + \varepsilon_X \cdot \% \text{ change in } [X] \tag{3}$$

The elasticity coefficient (ε_x), with respect to the concentration of a metabolite X that affects the enzyme rate, measures the dependence of that rate on the concentration of that variable:

$$\varepsilon_X \cong \left(\frac{\% \text{ change in } v_i}{\% \text{ change in } [X]} \right)_{all\ other\ factors\ unchanged} \tag{4}$$

For the ratios in Equations 1 and 4 to assume magnitudes that are independent of the magnitude of the causative change in [X], the percentages should be small, say 1% or, preferably for the mathematics, even smaller. In experimental practice one needs a compromise and works with changes that are measurable in terms of signal to noise ratio. The mathematical expression for the ratio of percentage changes for such very small changes equals the relative derivative, which in turns equals the logarithmic derivative:

$$\varepsilon_X = \left(\frac{\dfrac{\partial v_i}{v_i}}{\dfrac{\partial [X]}{[X]}} \right)_{S\&e_i\ constant} = \left(\frac{\partial \log v_i}{\partial \log [X]} \right)_{S\&e_i\ constant} \tag{5}$$

The curved d (∂) indicates a so-called partial derivative, that is, the effect is considered of a change in one variable at the time. Log can refer to 10-based, 2-based or e-based logarithm. In the latter case it is written as 'ln'. In similarly precise mathematical terms Equation 3 then reads (Sauro, 1990):

$$d \ln v_i = d \ln e_i + \varepsilon_S \cdot d \ln [S] + \varepsilon_X \cdot d \ln [X] \tag{6}$$

Dividing both sides of this equation by the relative change in the rate v_i, one obtains:

$$1 = \frac{d \ln e_i}{d \ln v_i} + \varepsilon_S^i \cdot \frac{d \ln [S]}{d \ln v_i} + \varepsilon_X^i \cdot \frac{d \ln [X]}{d \ln v_i} \tag{7}$$

$d\ln v_i$ refers to the actual change in rate of the enzyme, not to the change in its activity (V_{max} or K_{cat}). Total regulation of the rate of process i is here written as a sum of regulation through change in the amount of enzyme, and changes in the metabolism around the enzyme: The latter two terms dissect the metabolic regulation into a part that depends on the changes in [S] and the part that depends on the changes in [X] (Sauro, 1990).

Ter Kuile and Westerhoff (2001) recognized that the former term on the right-hand side of Equation 7 represents the regulation of the reaction rate that proceeds through gene expression, gene expression typically adjusting the amount of enzyme. The other two terms would correspond to regulation by the immediate metabolic environment of the enzyme and would hence correspond to the metabolic regulation of the process (even though these changes themselves might ultimately result from gene-expression regulation of a number of enzymes in the pathway). Accordingly, Ter Kuile and Westerhoff wrote Equation 7 with two terms on the right-hand side:

$$1 = \rho_h + \rho_m \tag{8}$$

With the definitions for the hierarchical regulation coefficient:

$$\rho_h = \frac{d\ln e_i}{d\ln v_i} \tag{9}$$

And for the metabolic regulation coefficient:

$$\rho_m = \varepsilon_S^i \cdot \frac{d\ln[S]}{d\ln v_i} + \varepsilon_X^i \cdot \frac{d\ln[X]}{d\ln v_i} \tag{10}$$

Ter Kuile and Westerhoff (2001) noted that the hierarchical regulation coefficient is a particularly operational property of the system that is being regulated: the relative change in enzyme concentration, $d\ln e_i$, can be measured as the change in maximal enzyme activity (V_{max}) in a cell-free extract of the preparation, whereas the relative change in flux through the process, $d\ln v_i$, should be equal to the relative change in flux that also occurred as a result of the regulation that the organism is engaging in. Thus the hierarchical regulation coefficient comprises all regulation that leads to a change in enzyme activity that survives the preparation of a cell-free extract, that is, both change in enzyme concentration and stable covalent modification such as results from most forms of signal transduction.

For less familiar aspects of regulation, this analysis serves as a classification, putting them either in the group of hierarchical or in the group of metabolic ('direct') regulation. Allosteric regulation, such as results from the non-covalent binding of a low-molecular-weight substance or from enzyme–enzyme association (unless that would reproduce in the *in vitro* assay of enzyme activity), would not be measured under ρ_h, hence would be classified as 'metabolic' or 'direct' regulation. This is perhaps a small price one pays for cutting a Gordian knot in regulation analysis.

Westerhoff and colleagues (Rossell *et al.*, 2005, 2006; Ter Kuile and Westerhoff, 2001) also emphasized that Equation 7 was not as general as possible, nor were the definitions of hierarchical and metabolic regulation. Equation 2 may be rewritten, as:

$$v_i = g(e_i) \cdot v(S, X) \tag{11}$$

The important point here is that the function g is a non-negative function of the enzyme concentration only, whereas the function v does not depend on the concentration of the enzyme and is therefore not directly a function of the expression of the gene responsible for the reaction i. The basis of this equation is the fact that the enzyme is a catalyst that can never change the direction of the reaction, whereas the concentrations of substrates and products can do so, and the point at which the sign changes should be independent of the enzyme (Westerhoff and Van Dam, 1987). The multiplication in Equation 11 is useful as the relative change in the total then equals the sum of the relative changes in the factors or, more precisely, the logarithm of a multiplication of two factors equals the sum of the logarithms of those factors.

Here we add the consideration that Equation 11 is valid at any point in time, that is:

$$v_i(t) = g(e_i(t), k_{cat}(t)) \cdot v(S(t), X(t), K_M(t)) \tag{12}$$

The function $e(t)$ indicates that the cell may regulate the process i by increasing the concentration of the enzyme e_i in time, obviously through enhancing the expression of the gene encoding enzyme i. $k_{cat}(t)$ refers the proportional regulation of both the forward and the reverse $k_{cat-}(t)$ by covalent modification subsequent to signal transduction; both these terms will be grouped under hierarchical regulation; their multiplication equals the better known $V_{max}(t)$. $K_M(t)$ indicates that the cell may increase the rate of the process by regulating (e.g. by phosphorylation or by expressing isoenzymes; Rossell et al., 2007) the Michaelis constants of the enzyme. [S(t)] and [X(t)] refer to the time-dependent concentrations of the metabolites that affect the enzyme. Alterations of the latter three will be grouped under 'metabolic regulation'.

We shall assume that at $t = 0$ a change in some external condition occurs, which induces the living system to regulate the rate of process i. $t = 0$ will represent the condition after the perturbation has occurred but before the living cell has begun its response. v_i may refer to the rate of a process that is measured 'off-line', such as a metabolic capacity the organism has and that it may regulate as a response to an external perturbation, even though that perturbation is again absent when the metabolic capacity is being assayed (e.g. see Rossell et al., 2005).

Taking the logarithm of the equation, taking the difference between $t = t$ and $t = 0$, and writing $\Delta \ln y(t)$ for $\ln\{y(t) - \ln y(0)\} = \ln\{y(t) / y(0)\}$ (for any y), one finds:

$$\Delta \ln \{v_i(t)\} = \Delta \ln g\{e_i(t), k_{cat}(t)\} + \Delta \ln v\{S(t), X(t), K_M(t)\} \tag{13}$$

Dividing both sides by the relative change in rate $\Delta \ln \{v_i(t)\}$, one obtains Equation 8 again, but now with more general definitions of the regulation coefficients (Bruggeman et al., 2006):

$$1 = \rho_h(t) + \rho_m(t) \tag{14}$$

$$\rho_h(t) = \frac{\Delta \ln g\{e_i(t), k_{cat}(t)\}}{\Delta \ln v_i(t)} \tag{15}$$

$$\rho_m(t) = \frac{\Delta \ln \upsilon\{S(t), X(t), K_M(t)\}}{\Delta \ln v_i(t)} \tag{16}$$

As alluded to above, the hierarchical regulation coefficient consists of a gene expression term and a signal transduction term (covalent modification term affecting k_{cat}). Experimentally one should check that enzyme V_{max} is proportional to the enzyme concentration. If so, then the third term in the equation that follows disappears:

$$\rho_h(t) = \frac{\Delta \ln e_i(t)}{\Delta \ln v_i(t)} + \frac{\Delta \ln k_{cat}(t)}{\Delta \ln v_i(t)} + \frac{\Delta \ln\left(g\{e_i(t), k_{cat}(t)\} / \{e_i(t) \cdot k_{cat}(t)\}\right)}{\Delta \ln v_i(t)} \tag{17}$$

And:

$$\rho_h(t) = \rho_g(t) + \rho_s(t) \tag{18}$$

The subscripts g and s refer to gene expression and signal transduction (the part thereof that acts on k_{cat}), respectively.

The metabolic regulation coefficient ρ_m cannot readily be subdivided into components, as the expression for the dependence of the enzyme catalysed reaction on substrates, products and allosteric modifiers cannot be written as a multiplication of independent factors. The gene-expression component of the hierarchical regulation coefficient can often be subdivided in terms that relate to regulation at the level of protein metabolism and regulation at the level of mRNA synthesis and degradation.

Note, for the purposes of this simple analysis, we neglect other processes that could affect the dynamics of gene expression in eukaryotes such as RNA-processing, RNA-transport and so on. In what follows we shall focus on this gene expression term. We shall assume that the mRNA that directs the synthesis of protein$_i$ is a minority among all mRNA that is being translated by the ribosome. This has the consequence that the rate of synthesis of protein$_i$ will vary proportionally with the concentration of its mRNA. This form of rate equation comes from a paradigm whereby the ribosomes and RNA polymerase act as catalysts and the different mRNAs are competing substrates/ligands. The rate of translation and transcription will therefore increase proportionally with the ribosome and RNA polymerase concentrations, respectively. We shall here assume an excess of charged tRNAs, as well as an excess of nucleoside tri-phosphates and a constant ATP/ADP ratio. Protein$_i$ and mRNA$_i$ will be minority species amongst all that are degraded by the RNases and proteases, such that protein$_i$ and mRNA$_i$ degradation will occur at rates that are proportional to their concentrations. With all these assumptions the time dependence of RNA_i and $protein_i$ should follow the following equations:

$$\frac{d\left[RNA_i\right]}{dt} = k_{transcription} \cdot \left[RNApolymerase\right] - k_{RNase} \cdot \frac{\left[RNA_i\right]}{\left[RNA_{tot}\right]} \cdot \frac{\left[RNA_{tot}\right]}{K_{RNA_{tot}} + \left[RNA_{tot}\right]} \quad (19)$$

$$\frac{d\left[protein_i\right]}{dt} = k_{translation} \cdot \left[RNA_i\right] \cdot \left[ribosome\right] - k_{protease} \cdot \frac{\left[protein_i\right]}{\left[protein_{tot}\right]}$$
$$\cdot \frac{\left[protein_{tot}\right]}{K_{protein_{tot}} + \left[protein_{tot}\right]} \quad (20)$$

For simplicity we shall assume that RNA metabolism is fast, such that the mRNA$_i$ concentration is in quasi-steady state (Heinrich *et al.*, 1977):

$$\left[RNA_i\right] = \frac{k_{transcription} \cdot \left[RNApolymerase\right] \cdot \left(K_{RNA_{tot}} + \left[RNA_{tot}\right]\right)}{k_{RNase}} \quad (21)$$

The protein concentration will vary with time according to:

$$\frac{protein_i(t)}{protein_i(0)} = 1 +$$
$$\left(\frac{\varphi\left(k_{translation}\right) \cdot \varphi\left[RNA_i\right] \cdot \varphi\left[ribosome\right] \cdot \varphi\left(K_{protein_{tot}} + \left[protein_{tot}\right]\right)}{\varphi\left(k_{protease}\right)} - 1 \right) \quad (22)$$
$$\cdot \left(1 - e^{-\frac{k_{protease} \cdot t}{K_{protein_{tot}} + \left[protein_{tot}\right]}} \right)$$

Here $\varphi(k_{translation})$ represents the factor by which the cell has up-regulated the translation rate constant at time zero:

$$\varphi(k_{translation}) = \frac{k_{translation\ after\ regulation}}{k_{translation\ before\ regulation}} \quad (23)$$

In the case of strong up-regulation, soon the above equation may be approximated by (we also take the logarithm of both sides and the difference between after and before regulation):

$$
\begin{aligned}
\Delta \ln(protein_i(t)) - \ln\left(1 - e^{-\frac{k_{protease} \cdot t}{K_{protein_{tot}} + [protein_{tot}]}}\right) \\
= \Delta \ln\left[RNA_i\right] + \Delta \ln\left(k_{translation}\right) + \Delta \ln\left[ribosome\right] - \Delta \ln\left(k_{protease}\right) \\
+ \Delta \ln\left(K_{protein_{tot}} + [protein_{tot}]\right)
\end{aligned}
\tag{24}
$$

We shall here assume that the cell up-regulates parameters at time zero to a new value which it then maintains. Gene expression regulation happened immediately after time = 0 and the mRNA level adjusted quickly to any such regulation at the level of transcription or mRNA stability. Accordingly, the regulation itself is not time dependent. Yet, because of the slower relaxation time of protein synthesis the instantaneous regulation at and above the protein level affects protein concentration and reaction rate v_i only slowly. This phenomenon is represented by the second, time-dependent, term on the left-hand side of Equation 24.

For the regulation of the protein level we now find that it is composed of a number of terms that add up to 100% regulation: We divide Equation 24 by the two terms on the left-hand side:

$$
1 = \rho_{p,mRNA} + \rho_{p,translation} + \rho_{p,[ribosome]} + \rho_{p,protease} + \rho_{p,protease\ competition}
\tag{25}
$$

These protein-level regulatory coefficients in this equation are defined by:

$$
\rho_{p,mRNA} = \frac{\Delta \ln\left[mRNA_i\right]}{\Delta \ln(protein_i(t)) - \ln\left(1 - e^{-\frac{k_{protease} \cdot t}{K_{protein_{tot}} + [protein_{tot}]}}\right)}
\tag{26}
$$

$$
\rho_{p,translation} = \frac{\Delta \ln k_{translation}}{\Delta \ln(protein_i(t)) - \ln\left(1 - e^{-\frac{k_{protease} \cdot t}{K_{protein_{tot}} + [protein_{tot}]}}\right)}
\tag{27}
$$

$$
\rho_{p,protease} = \frac{-\Delta \ln k_{protease}}{\Delta \ln(protein_i(t)) - \ln\left(1 - e^{-\frac{k_{protease} \cdot t}{K_{protein_{tot}} + [protein_{tot}]}}\right)}
\tag{28}
$$

$$P_{p,protease\ competition} = \frac{\Delta \ln\left(K_{protein_{tot}} + \left[protein_{tot}\right]\right)}{\Delta \ln(protein_i(t)) - \ln\left(1 - e^{\frac{k_{protease}\cdot t}{K_{protein_{tot}} + \left[protein_{tot}\right]}}\right)} \tag{29}$$

The regulation by protease is negative if the protease is activated. The regulation coefficient by protease competition can be positive or negative; it reflects the potential effect that if total protein level in the cell decreases (or increases) at constant level of protein$_i$, proteolysis of protein$_i$ increases (or decreases) simply because of less competition with other proteins for the protease. The other regulation coefficients are positive when the corresponding process is activated by the cell in its regulatory activity that leads to an increase in the rate v_i.

Using Equation 21, also the regulation at the mRNA level can be subdivided into a number of potential regulations that add up to a total of 1:

$$1 = P_{R,transcription} + P_{R,RNA\ polymerase} + P_{R,RNase} + P_{R,RNase\ competition} \tag{30}$$

The corresponding definitions are:

$$P_{R,transcription} = \frac{\Delta \ln k_{transcription}}{\Delta \ln[RNA_i]} \tag{31}$$

$$P_{R,transcription} = \frac{\Delta \ln\left[RNApolymerase\right]}{\Delta \ln[RNA_i]} \tag{32}$$

$$P_{R,RNase} = \frac{\Delta \ln k_{RNAase}}{\Delta \ln[RNA_i]} \tag{33}$$

$$P_{R,RNase\ competition} = \frac{\left(K_{RNA_{tot}} + \left[RNA_{tot}\right]\right)}{\Delta \ln[RNA_i]} \tag{34}$$

3 Hierarchical Regulation Analysis: A practical approach

In the preceding section the conceptual and theoretical basis of Hierarchical Regulation Analysis (HRA) was reviewed and extended. This showed that the approach is based on the known organization of cellular biochemistry and molecular biology. It constitutes the elaboration of the 'dogma of molecular biology' to a quantitative analysis method. It enables one to determine unequivocally and experimentally where a certain intracellular process is regulated. It shows how to determine whether, when the organism is confronted with a certain challenge, it regulates a process of interest metabolically or through gene expression. It also shows the extent to which total regulation resides within each of these categories. We shall now make explicit how the analysis is actually performed.

First, one should of course decide in which case of regulation by the organism one is interested. This means that one should define of which process one wishes to study the regulation and as a response to which challenge. The requirement that one needs to specify both the process and the challenge has become clear in earlier regulation analysis: When Rossell *et al.* (2006) studied the regulation of the steady-state rates of glycolytic enzymes in yeast, the distribution between metabolic and gene-expression regulation they observed differed between various enzymes in the glycolytic pathway. Moreover, this distribution differed depending on whether the cells were responding to carbon starvation or to nitrogen starvation. This very result was of historical significance, as earlier many studies on regulation suggested they had found answers of general validity, such as that phosphofructokinase was the regulator of the glycolytic flux (the statement implying that there was only one flux to be regulated and that that could be accomplished by a single enzyme, that the answer should always be the same, and that the extent of regulation should always be the same, i.e. complete). HRA shows that all the implicit assumptions were inappropriate.

A simple question one may ask then is: What is the extent to which the cell regulates an intracellular process when adapting its steady-state functioning to the challenge? One is then interested in the regulation of the change in the rate of a process i between two steady states. The first thing to do, then, is to measure to an accuracy of 20% or more, the change the steady-state fluxes through the process before and after the challenge. This can be done in arbitrary units. One then takes the logarithm (10-based, e-based, or 2-based; this does not matter as long as one is consistent) of each rate and the difference between the two logarithms. For instance, if the cells were suddenly incubated at a different pH, and responded by increasing (among others) the rate of process 3 from 2 nmoles min^{-1} $mgprotein^{-1}$ to 8 nmoles min^{-1} $mgprotein^{-1}$, then:

$$\Delta \log_2 v_3 = \log_2 v_3 (final) - \log_2 v_3 (begin) = \log_2 v_3 (8) - \log_2 v_3 (2) = 3 - 1 = 2 \qquad (35)$$

Then one should prepare cell extracts and measure the activity in terms of V_{max} of the enzyme catalysing reaction 3. Again, the units of activity do not matter, and neither does the efficiency of preparing the cell-free extract as long as the efficiency is the same before and after the regulation. Indeed one could do special, mixed or spiked assays of the two extracts so as to ensure that these conditions are met. Let us assume that the V_{max} of the cell-free extract after the regulation was twice as high as the V_{max} of the extract of the cells before the regulation. We assume that V_{max} changes are the result of changes in enzyme concentration:

$$\Delta \log_2 e_3 = \Delta \log_2 2 = 1 \qquad (36)$$

Of the four-fold change in rate (i.e. $8/2 = 4$), two fold is then explained by the change in enzyme concentration, leaving another two-fold change to be explained by metabolic regulation, that is:

$$\rho_b^3 = \frac{\Delta \log_2 e_3}{\Delta \log_2 v_3} = \frac{1}{2} = 0.5 \tag{37}$$

$$\rho_m^3 = 1 - 0.5 = 0.5 \tag{38}$$

In words, in this example 50% of the regulation of the flux through process 3 was through gene expression and 50% was through metabolic interactions.

We assume that in this example the regulation is studied between two long-term steady states, where the protein level has also reached steady state. Let us further assume that the mRNA corresponding to the protein 3 was measured and found to be increased by a factor of 1.4 and that the ribosome concentration and the protein$_3$ and mRNA$_3$ stability were found to be constant. The conclusions should be that all regulation at the mRNA$_3$ level was at the level of transcription and that the regulation at the protein level was distributed between the dosage of mRNA and the translation rate constant:

$$\rho_{p,RNA} = \frac{\log_2 1.4}{\log_2 2} = 0.5 \tag{39}$$

$$\rho_{p,k_{translation}} = 1 - 0.5 - 0 - 0 - 0 = 0.5 \tag{40}$$

$$\rho_{R,k_{transcription}} = \frac{\log_2 1.4}{\log_2 1.4} = 1 \tag{41}$$

The grand total would be that the flux through process 3 was regulated 50% metabolically, 25% at the level of translation activity and 25% at transcription.

All that were needed to reach these conclusions were experiments that were straightforward with present-day techniques, even though they require more than usual accuracy and persistence. No mathematics and few assumptions were needed really. The mathematics used in the previous section merely served to give and delimit solid foundations of what is basically intuitively clear.

We shall now give some results of an ongoing experimental study, and then present an *in silico* simulation suggesting what the type of basis might be for a diversification of regulation as observed here.

3.1 *An exemplary experimental analysis: Alcohol production by yeast as a function of time after nitrogen starvation*

In an ongoing experimental study in baker's yeast, we aim to determine the regulation of the rates of various glycolytic steps as a function of time after the cells have been subjected to nitrogen starvation. In this case it is not the actual flux through a step that is being regulated, but the potential of the cells to engage in such a flux when subsequently confronted with excess glucose, inducing them to ferment maximally. The assay is relevant to the culture of yeast that is to be used for the baking of bread, where the main function is to produce as much carbon dioxide as possible in a period of slightly over an hour, but may also be useful for certain processes producing alcohol.

In an earlier study some of us had determined the regulation after nitrogen starvation as effective in the new steady state, which we assumed to have been attained 24 hours after the onset of the starvation (Rossell *et al.*, 2005). Here we were interested in the development of this regulation as a function of time. At least one of us (HVW) had the working hypothesis that the initial regulation would be entirely metabolic, essentially because there would not have been sufficient time for gene expression regulation to have had an effect at the protein level. He expected gene expression regulation to become important only subsequently and then to become dominant, making metabolic regulations disappear.

Implementing HRA we first established the process we wanted to study the regulation of (*Figure 3*): This was to be the capability of yeast to produce alcohol, that is, the flux through alcohol dehydrogenase. We also decided which regulation of this

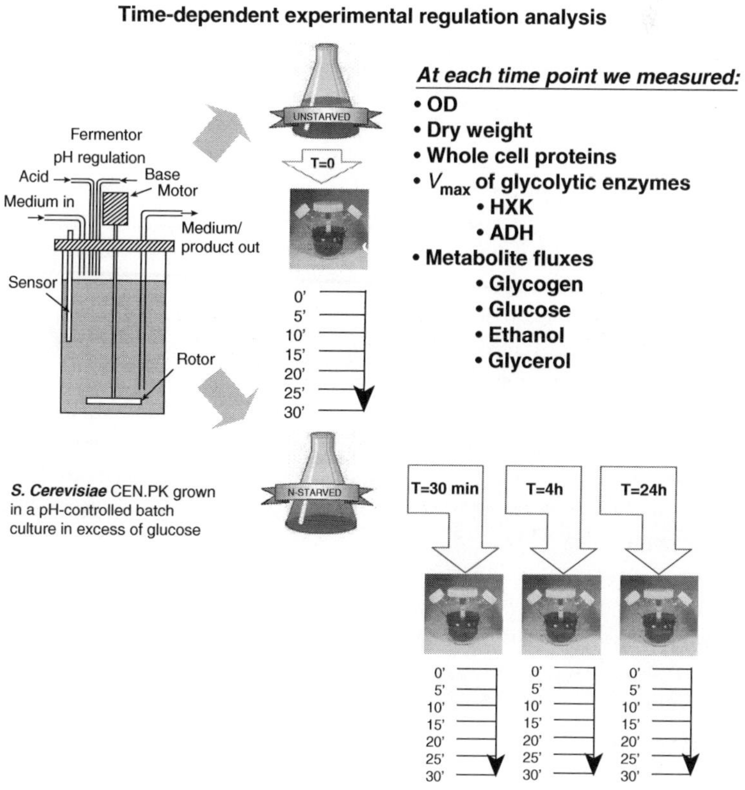

Figure 3. Schema of the experiment performed to determine the regulation of the flow through the glucose phosphorylation and the alcohol production step in yeast under fermentative conditions, the regulation occurring after the cells were starved for nitrogen.

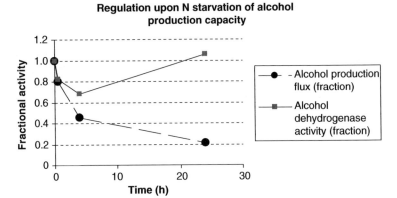

Figure 4. *Variation of alcohol production flux and of alcohol dehydrogenase activity with time after initiation of nitrogen starvation in yeast. S. cerevisiae was cultivated as described in Figure 3 and then subjected for the indicated time to incubation in the absence of any external nitrogen source. At the indicated time points, the cells were harvested and washed and then given excess glucose to be tested for alcohol production capability under fully fermentative conditions.*

process we wished to study: The regulation the cell undertakes after having been confronted with a shortage of nitrogen. The expectation was that perhaps as a result of the nitrogen shortage, the cell would soon engage in major down-regulation of the expression level of the enzymes it needed for its fermentative capacity, such as alcohol dehydrogenase. We therefore first grew the cells under steady nitrogen, carbon and Gibbs free energy excess conditions in order to bring them into a reproducible 'affluent' state. We then washed the cells and incubated them in the absence of any external nitrogen source for various lengths of time. Harvesting the cells after this starvation period we then incubated an aliquot under our standard fermentative capacity assay conditions and measured the rate at which alcohol appeared in the medium. Other aliquots were used to prepare cell-free extracts, in which we measured the V_{max} of alcohol dehydrogenase activity. For the essence of the experimental procedures we refer the reader to Rossell *et al.* (2006), with a few modifications (Bevilacqua, unpublished results).

The circles in *Figure 4* show how indeed the cells down-regulated their alcohol production capacity after they had been confronted with the absence of any nitrogen source. After 24 hours only 20% of the capacity was left. Initially this was accompanied by a rather strong decrease in alcohol dehydrogenase concentration (or at least measured V_{max} in the extract: The squares in *Figure 4*). As shown in *Figure 5*, the extent of this initial down-regulation of the amount of alcohol dehydrogenase proved the expectation of one of us to be false: There was no evidence of a dominance of metabolic regulation early on, slowly giving way to hierarchical regulation. Rather the opposite was observed for alcohol dehydrogenase: its hierachical regulation coefficient started out close to 1 and then dwindled to zero. Accordingly the metabolic regulation coefficients started out quite low and then increased to close to 1 at 24 hours into starvation.

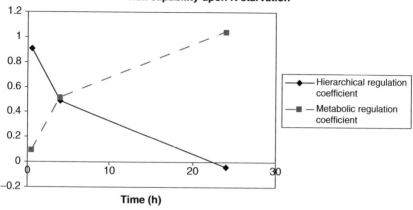

Figure 5. *Hierarchical and metabolic regulation of alcohol production capability of* S. cerevisiae *when confronted with nitrogen depletion.*

Table 1 shows the precise details of all the calculations and might serve as a paradigm for others engaging in the study of hierarchical regulation. Its first three columns show the time and the two experimental observables. The next two columns show how we converted the measured activities into fractions of the initial value. The next two column shows what we obtained by taking the logarithms of these values (we took e-based logarithms, but one may also take 10- or 2-based logarithms). The subsequent column shows how we calculated the hierarchical regulation coefficient by dividing the logarithm of the fractional alcohol dehydrogenase activity by the logarithms of the fractional alcohol production flux. The final column shows how we obtained the metabolic regulation coefficients by subtracting the hierarchical regulation coefficient from 1.

Figure 5 reinforces that regulation of an important function of a living organism is not just regulated either metabolically or by gene expression alone. Rather it is a mixture of both where the composition of the mixture may vary with time. This variation can be counter-intuitive in the sense that gene expression regulation precedes metabolic regulation, where the latter may surprisingly be the one persisting near the steady state. On the basis of earlier work (Rossell *et al.*, 2005; Ter Kuile and Westerhoff, 2001) we note that the particular regulation of the alcohol production flux observed here is not representative of the regulation of other processes in the organism and not even of the regulation of alcohol production flux under all possible experimental conditions: at the moment it seems that regulation is a rich phenomenon that living organisms engage in. We would speculate that this is because different organisms are different organisms precisely because they have to be fit under different circumstances using different processes optimally and making different sets of processes and concentrations optimal. Viewing this complexity of regulation, one is tempted to be suspicious of single fluxes as objective functions in flux balance analysis (Palsson, 2006).

Table 1. Procedure of the calculation of the hierarchical and metabolic regulation coefficients.

Time (h)	Ethanol flux (mmol min⁻¹ gprotein⁻¹)	Alcohol dehydrogenase activity (mmol min⁻¹ gprotein⁻¹)	Ethanol flux (fraction)	Alcohol dehydrogenase activity (fraction)	ln(ethanol flux fraction)	ln(adh activity fraction)	$\rho_{hierarchical}$	$\rho_{metabolic}$
T	$J(t)$ From experiment	$A(t)$ From experiment	$FE = J(t)/J(0)$	$FA = A(t)/A(0)$	$LE = {}^{e}\log(FE)$	$LA = {}^{e}\log(FA)$	$RHOh = LA/LE$	$1-RHOh$
0	0.828	4.7	1	1	0	0		
0.5	0.663	3.87	0.8	0.82	−0.22	−0.2	0.91	0.09
4	0.374	3.21	0.45	0.68	−0.8	−0.39	0.49	0.51
24	0.176	4.99	0.21	1.06	−1.56	0.06	−0.04	1.04

Time after onset of nitrogen starvation, ethanol production flux and alcohol dehydrogenase activity were determined experimentally. All other values were calculated from these results as indicated on the second row. ln refers to e-based logarithm, which may be replaced by 10-based or 2-based logarithms provided this is done consistently. We emphasize that these data stem from only two experiments with the V_{max} assays having been done in duplicate and triplicate only. Variations between strains, preincubation conditions and time points are still under investigation.

3.1.1 An illustration *in silico*: Yeast glycolysis

Much of the regulation behaviour that has been discovered with HRA was unexpected and it has been difficult to predict intuitively what type of regulation might occur and when. In this section we shall therefore present a regulatory network that is partly known and partly extended by reasonable, yet speculative, regulatory interactions (see also Bruggeman *et al.*, 2006). We shall then analyse what the consequences are of the knowledge and speculate on the distribution of the regulation. To do this we shall focus on the hierarchical regulation of a single process (glucose transport) and adapt an existing mathematical model of yeast glycolysis (Teusink *et al.* 2000) in order to take account of the results of Rossell *et al.* (2005, 2006).

We model the transcription and translation of the glycolytic enzymes assuming irreversible reactions for the synthesis and degradation of the enzymatic RNA and the enzyme:

$$DNA_j + RNAPolymerase + RNANucleotides \xrightarrow{k_j^{RNASynthesis}} RNA_j + DNA_j + RNAPolymerase$$

$$RNA_j \xrightarrow{k_j^{RNADegradation}} RNANucleotides$$

$$RNA_j + Ribosome + AminoAcids \xrightarrow{k_j^{EnzymeSynthesis}} Enzyme_j + RNA_j + Ribosome$$

$$Enzyme_j \xrightarrow{k_j^{EnzymeDegradation}} AminoAcids$$

Since the mechanisms of the suspected (Rossell *et al.*, 2006) regulation of translation by metabolic intermediates, $(X_1, X_2, .., X_n)$, are not well understood we shall model their effect by introducing the function $f_j^{EnzymeSynthesis}(X_1, X_2, .., X_n)$. We can then write the equations for the rates of change of the enzymatic RNA and the enzyme as (note in this case we explicitly include DNA in the rate of expression in order to be consistent with the reaction scheme given above):

$$\frac{d}{dt}\left[RNA_j\right] = k_j^{RNASynthesis} \cdot \left[DNA_j\right] \cdot \left[RNAPolymerase\right] \cdot \left[RNANucleosidesTP\right]$$
$$- k_j^{RNADegradation} \cdot \left[RNA_j\right] \qquad (42)$$

$$\frac{d}{dt}\left[Enzyme_j\right] = k_j^{EnzymeSynthesis} \cdot f_j^{EnzymeSynthesis}(x_1, x_2, .., x_n)$$
$$\cdot \left[RNA_j\right] \cdot \left[Ribosome\right] \cdot \left[AminoAcids\right] \qquad (43)$$
$$- k_j^{EnzymeDegradation} \cdot \left[Enzyme_j\right]$$

We shall consider the hierarchical regulation of a single enzyme – the glucose transporter (GLT), which may well be the step that exerts most control over the pathway flux (Reijenga *et al.*, 2001). We consider a simple instance of the above scheme, assuming that transcriptional flux is not a function of any metabolite concentrations and that the 'concentrations' of DNA, RNA polymerase and the RNA nucleoside-tri phosphates are approximately constant and can therefore be 'lumped' into a single effective rate constant $\hat{k}_{GLT}^{RNASynthesis}$. For the translational flux we assume that metabolic influence is restricted to pyruvate (PYR). We present this merely as a plausible effect since pyruvate is known to be a key precursor in synthesis of amino acids (Golshani-Hebroni and Bessman, 1997). We model this influence with saturable kinetics using a Michaelis–Menten function and assuming a K_m value equal to the steady value of pyruvate in the simulation of the adjusted model (*Table 2*). So, in this case, we have:

$f_{GLT}^{EnzymeSynthesis}\left(X_1, X_2, ..., X_n\right) = \dfrac{2 \cdot x_{PYR}}{14.1 + x_{PYR}}$. We perform a similar lumping as for the

transcriptional case whereby the concentrations of the ribosomes and amino acid pool are taken to be constant and lumped into a single rate constant $\hat{k}_{GLT}^{EnzymeSynthesis}$:

$$\frac{d}{dt}\left[RNA_{GLT}\right] = \hat{k}_{GLT}^{RNASynthesis} - k_{GLT}^{RNADegradation} \cdot \left[RNA_{GLT}\right] \tag{44}$$

$$\frac{d}{dt}\left[Enzyme_{GLT}\right] = \hat{k}_{GLT}^{EnzymeSynthesis} \cdot \frac{2x_{PYR}}{14.1 + x_{PYR}} \cdot \left[RNA_{GLT}\right]$$
$$- k_{GLT}^{EnzymeDegradation} \cdot \left[Enzyme_{GLT}\right] \tag{45}$$

We estimated values for the synthesis rate constants above by assuming reasonable values for the steady-state concentrations of RNA and enzymes and also for their rates of degradation (*Table 2*).

Table 2. *Estimated and calculated model parameters.*

Yeast cell volume	4×10^{-14} litre
Steady state concentration of RNA_{GLT} (based on 10^2 molecules per cell)	4.15 nM
Steady state concentration of $Enzyme_{GLT}$ (based on 10^5 molecules per cell)	4.15 μM
Steady state concentration of pyruvate x_{PYR} based on simulation of adjusted Teusink model)	14.6 mM
$k_{GLT}^{RNADegradation}$ (based on half-life of 20 minutes)	0.0347 min^{-1}
$k_{GLT}^{EnzymeDegradation}$ (based on half-life of 60 minutes)	0.0116 min^{-1}
$\hat{k}_{GLT}^{RNASynthesis}$ (calculated from Eq. 44) assuming steady state − left-hand side = zero)	1.44×10^{-7} mM min^{-1}
$\hat{k}_{GLT}^{EnzymeSynthesis}$ (calculated from Eq. 45) assuming steady state − left-hand side = zero)	11.6 min^{-1}

These values correspond to initial estimates and will be optimized in future work. They should not be used elsewhere.

For the metabolic part, our mathematical model is based on the yeast glycolysis model of Teusink *et al.* (2000). We have *adjusted* the core model in order to take account of the results obtained by Rossell *et al.* (2006), who reported zero flux in the trehalose and glycogen branches for yeast cells grown under glucose-starved and unstarved conditions, respectively. We have therefore fixed these fluxes to zero in our model. We have also augmented the model to include the hierarchical regulation of the glucose transporter as described above. For the glucose transport flux we have replaced the Teusink V_{GLT}^{max} parameter with $k_{GLT}^{cat} \cdot \left[Enzyme_{GLT} \right]$ in order to introduce the dependence of glucose transport flux on the concentration of the glucose trans-porter enzyme. We calculated the value of k_{GLT}^{cat} by dividing the Teusink V_{GLT}^{max} para-meter by the assumed steady-state concentration of enzyme (*Table 2*):

$$k_{GLT}^{cat} = \frac{V_{GLT}^{max}}{\left[Enzyme_{GLT} \right]_{SS}} = \frac{97.3}{0.00415} = 2.34 \times 10^4 \text{ min}^{-1} \tag{46}$$

Finally, we introduced Equations 44 and 45 to describe the time evolution of the glu-cose transporter enzyme and RNA.

We simulated the adapted model using a fixed concentration of extracellular glucose $GLU_o = 50$ mM and found a steady state for all metabolites and for both the glucose transporter RNA and protein concentrations. The time taken to reach steady state was much longer than 1 hour, whereas the original Teusink *et al.* (2000) model took under 1 minute. The introduction of the simple transcriptional terms for the glucose transporter with the long half life of the glucose transport protein, slowed down the dynamics of the system by several orders of magnitude.

In order to investigate the hierarchical control during glucose starvation we com-pared simulations for different values of GLU_o (50 mM *versus* 5 mM). The results are shown in *Table 3*. We then calculated the hierarchical and metabolic regulation coefficients for glucose transport subjected to this simulated glucose starvation. The glucose transporter flux was:

$$v_{GLT} = k_{GLT}^{cat} \left[Enzyme_{GLT} \right] \cdot v_{GLT}(x_1, x_2, ..., x_n) \tag{47}$$

For the glucose transporter the general rate equation is a function of the extra-cellular and intra-cellular glucose concentrations only so that we write:

$$v_{GLT} = k_{GLT}^{cat} \left[Enzyme_{GLT} \right] \cdot v_{GLT}(x_{GLT_o}, x_{GLT_i}) \tag{48}$$

Substituting values from *Table 3* for the corresponding variables, we calculated the value for the hierarchical control coefficient.

$$\rho_h = \frac{\ln\left(\dfrac{1.12}{0.815} \right)}{\ln\left(\dfrac{25.0}{15.0} \right)} = 0.62 \tag{49}$$

Table 3. *Steady-state values calculated for the mathematical model (see text)*

Extracellular glucose concentration GLU_o (mM)	Flux through glucose transporter v_{GLT} (mM min^{-1})	Pyruvate concentration x_{PYR} (mM)	Glucose transporter RNA concentration $[RNA_{GLT}]$ (nM)	Glucose transporter enzyme concentration $[Enzyme_{GLT}]$ (μM)
50	25.0	2.28	4.15	1.12
5	15.0	1.59	4.15	0.815

This value is entirely due to translational control since our assumption of a constant transcriptional rate implies zero transcriptional control. The summation theorem gave the metabolic regulation coefficient:

$$\rho_m = 1 - \rho_h = 0.38 \tag{50}$$

In this case, the hierarchical and metabolic regulation work in the same direction, the hierarchical regulation being dominant. In other words, the reduced enzyme levels (indirect effect of glucose starvation) have a bigger effect in the flux decrease than the direct effect of the reduced extra-cellular glucose concentration in the glucose transporter rate equation.

This simplistic result partially agrees with Rossell *et al.* (2006), who also found hierarchical regulation is synergistic with metabolic regulation for the glucose transporter under carbon starvation. However, they found that metabolic regulation was stronger ($\rho_h = 0.4$, $\rho_m = 0.6$).

4 Concluding remarks

There are many practical approaches to systems biology. Many of these observe correlations between the changes in concentrations of macromolecules. By mapping these onto maps of the intracellular networks, inferences are sought about co-regulation of those pathways under the conditions that are being studied. All too often these types of analysis inspect only one level of the cellular organization, be it the level of mRNA, the level of protein, or the level of metabolites. The ability to decipher regulatory or even flux pathways in this manner therefore depends on either (i) all or most regulation involving only one of these levels and indeed the level that is being investigated, or (ii) all levels of cellular organization being co-regulated proportionally.

In the proportional regulation paradigm proposed by Small and Kacser (1994) and Fell (2005), flux through a metabolic pathway should be regulated through the simultaneous and proportional alteration in the level of all its enzymes. Then there should be no change in the concentrations of the pathway intermediates with the advantage that all other cellular processes that may depend on these intermediates are not dysregulated when the prime process is being regulated. The simple extension of this paradigm to the RNA world is that transcription is also up-regulated proportionally with enzyme activation.

Attractive as this proportional regulation paradigm might be, it can only serve as a possibility, the reality of which needs to be examined experimentally. It should require that all hierarchical regulation coefficients equal 1, all metabolic regulation coefficients equal 0, the mRNA regulation coefficient at the protein synthesis level equals 1 and the transcription regulation for the mRNAs all equal 1.

HRA enables the testing of the paradigm. Thus far, application of HRA has shown that neither yeast nor trypanosomes make use of the proportional regulation paradigm when regulating the fluxes through their glycolytic pathway in response to carbon substrate or nitrogen limitation: For many enzymes metabolic regulation coefficients were much above zero (cf. *Figure 5*); they differed between the enzymes in the same pathway; the mRNA coefficient for regulation of protein synthesis was often much below 1 (Daran-Lapujade *et al.*, 2007). In these cases the organisms did not implement the proportional regulation paradigm.

This should not detract from the immense value of the concept of proportional regulation. There may well be cases where it applies. It does bring home, however, that there is much more to be learned about the 'rationale' organisms use when 'choosing' their regulatory strategy.

The HRA results obtained thus far may compromise, however, the mapping of regulatory pathways that is being practised on the basis of measuring only a single level of cellular organization. There is no basis for such an activity and had it been applied to yeast glycolysis, a pathway would have resulted with many steps missing and with additional steps being inserted. The problem is that most of these systems biology approaches that have been implemented thus far are not integrative enough: they deal with integration at a single level of organization (such as the transcriptome), constructing interactions that may not be present at that level, but neglecting interactions that involve other levels of the cellular control hierarchy.

The approach we delineated in this chapter addresses this problem head on: for a cellular function of interest it determines whether it is solely regulated at any of the levels of the cellular control hierarchy. If it is, then this rationalizes one of the existing systems biology approaches, namely the one studying the relevant level. Of course, it then simultaneously devalues the systems biology approaches limited to any of the other levels, but only for that particular case of regulation. If it is not regulated at a single level of the cellular regulation hierarchy, then HCA determines quantitatively how much of each level is involved. It also suggests how then each level should be analysed in turn. This additional analysis does not correspond to what is usually done, for example determining the patterns of variation of the genome-wide transcriptome. It rather focuses on the regulation of the metabolism of the most relevant mRNA, that is on transcription, RNA polymerase concentration, mRNA stability, and so on.

This Hierarchical Regulation Analysis is a practical approach, in that it largely depends on, and consists of, experimental determination of the variation of intracellular function (e.g. flux through a process), the corresponding enzyme activity, the corresponding protein concentration and the corresponding mRNA abundance with the progression of the regulation. The theory around it serves to justify the approach but is not used when actually applying it, as should be clear from our experimental example and *Table 1*.

This is not to say that more theoretical and modelling approaches are not useful as well (Westerhoff, 2007). They may enable us to establish mechanisms for the emergence of the new regulatory properties and determine how the emergent properties depend on the various processes in the system.

Acknowledgements

We thank various research funding organizations for support, including BBSRC, EPSRC, NWO, STW, NGN, and EC-FP6 (BioSim, NucSys, YSBN, EC-MOAN).

References

Alberghina, L. & Westerhoff, H.V. (eds.) (2005) *Systems Biology: Definitions and Perspectives*. Springer, Berlin.

Boogerd, F.C., Bruggeman, F.J., Hofmeyr, J.H.S. & Westerhoff, H.V. (eds.) (2007) *Systems Biology: Philosophical Foundations*. Elsevier, Amsterdam.

Bruggeman, F.J., de Haan, J., Hardin, H., Bouwman, J., Rossell, S., van Eunen, K., Bakker, B.M. & Westerhoff, H.V. (2006) Time-dependent hierarchical regulation analysis: deciphering cellular adaptation. *IEE Proc. Syst. Biol.* **153:** 318–322.

Burns, J.A., Cornish-Bowden, A., Groen, A.K. *et al.* (1985) Control analysis of metabolic systems. *Trends Biochem. Sci.* **10:** 16.

Daran-Lapujade, P., Rossell, S., van Gulik, W.M. *et al.* (2007) The fluxes through glycolytic enzymes in *S. cerevisiae* are predominantly regulated at posttranscriptional levels. *Proc. Natl Acad. Sci. USA* **104:** 15753-15758.

Even, S., Lindley, N.D. & Cocaign-Bousquet, M. (2003) Transcriptional, translational and metabolic regulation of glycolysis in *Lactococcus lactis* subsp. *cremoris* MG 1363 grown in continuous acidic cultures. *Microbiology* **149:** 1935–1944.

Fell, D.A. (2005) Enzymes, metabolites and fluxes. *J. Exp. Bot.* **56:** 267–272.

Golshani-Hebroni, S.G. & Bessman, S.P. (1997) Hexokinase binding to mitochondria: a basis for proliferative energy metabolism. *J. Bioenerg. Biomembr.* **29:** 331-338.

Groen, A.K., Wanders, R.J.A., Westerhoff, H.V., Van der Meer, R. & Tager, J.M. (1982) Quantification of the contribution of various steps to the control of mitochondrial respiration. *J. Biol. Chem.* **257:** 2754–2757.

Heinrich, R. & Rapoport, T.A. (1974) A linear steady-state treatment of enzymatic chains. General properties, control and effector strength. *Eur. J. Biochem.* **42:** 89–95.

Heinrich, R., Rapoport, T.A. & Rapoport, S.M. (1977) Metabolic regulation and mathematical models. *Progr. Biophys. Mol. Biol.* **32:** 1–82.

Hofmeyr, J.H. & Westerhoff, H.V. (2001) Building the cellular puzzle: control in multi-level reaction networks. *J. Theor. Biol.* **208:** 261–285.

Kacser, H. & Burns, J.A. (1973) Control of flux. *Symp. Soc. Exp. Biol.* **27:** 65–104.

Kahn, D. & Westerhoff, H.V. (1991) Control theory of regulatory cascades. *J. Theor. Biol.* **153:** 255–285.

Middleton, R.J. & Kacser, H. (1983) Enzyme variation, metabolic flux and fitness: alcohol dehydrogenase in *Drosophila melanogaster*. *Genetics* **105:** 633–650.

Palsson, B.O. (2006) *Systems Biology*. Cambridge University Press, Cambridge.

Reijenga, K.A., Snoep, J.L., Diderich, J.A., Van Verseveld, H.W., Westerhoff, H.V. & Teusink, B. (2001) Control of glycolytic dynamics by hexose transport in *Saccharomyces cerevisiae*. *Biophys. J.* **80:** 626–634.

Rossell, S., van der Weijden, C.C., Kruckeberg, A.L., Bakker, B.M. & Westerhoff, H.V. (2005) Hierarchical and metabolic regulation of glucose influx in starved *Saccharomyces cerevisiae*. *FEMS Yeast Res.* **5**: 611–619.

Rossell, S., van der Weijden, C.C., Lindenbergh, A., van Tuijl, A., Francke, C., Bakker, B.M. & Westerhoff, H.V. (2006) Unraveling the complexity of flux regulation: a new method demonstrated for nutrient starvation in *Saccharomyces cerevisiae*. *Proc. Natl Acad. Sci. USA* **103**: 2166–2171.

Rossell, S., Lindenbergh, A., van der Weijden, C.C., Kruckeberg, A.L., van Eunen, K., Westerhoff, H.V. & Bakker, B.M. (2007) Mixed and diverse metabolic and gene-expression regulation of the glycolytic and fermentative pathways in response to a *HXK2* deletion in *Saccharomyces cerevisiae*. *FEMS Yeast Res.* (OnlineEarly Articles) doi:10.1111/j.1567-1364.2007.00282.x

Sauro, H.M. (1990) Quantification of metabolic regulation by effectors. In: *Control of Metabolic Processes* (eds A. Cornish-Bowden and M.L. Cardenas). NATO ASI Series, Plenum Press, New York, pp. 225–230.

Small, J.R. & Kacser, H. (1994) A method for increasing the concentration of a specific internal metabolite in steady-state systems. *Eur. J. Biochem.* **226**: 649–656.

Snoep, J.L., van der Weijden, C.C., Andersen, H.W., Westerhoff, H.V. & Jensen, P.R. (2002) DNA supercoiling in *Escherichia coli* is under tight and subtle homeostatic control, involving gene-expression and metabolic regulation of both topoisomerase I and DNA gyrase. *Eur. J. Biochem.* **269**: 1662–1669.

Suarez, R.K., Darveau, C.A. & Hochachka, P.W. (2005) Roles of hierarchical and metabolic regulation in the allometric scaling of metabolism in Panamanian orchid bees. *J. Exp. Biol.* **208**: 3603–3607.

Ter Kuile, B.H. & Westerhoff, H.V. (2001) Transcriptome meets metabolome: hierarchical and metabolic regulation of the glycolytic pathway. *FEBS Lett.* **500**: 169–171.

Teusink, B., Passarge, J., Reijenga, C.A. *et al.* (2000) Can yeast glycolysis be understood in terms of *in vitro* kinetics of the constituent enzymes? Testing biochemistry. *Eur. J. Biochem.* **267**: 5313–5329.

van der Gugten, A.A. & Westerhoff, H.V. (1997) Internal regulation of a modular system: the different faces of internal control. *Biosystems* **44**: 79–106.

Wanders, R.J., Groen, A.K., Van Roermund, C.W. & Tager, J.M. (1984) Factors determining the relative contribution of the adenine-nucleotide translocator and the ADP-regenerating system to the control of oxidative phosphorylation in isolated rat-liver mitochondria. *Eur. J. Biochem.* **142**: 417–424.

Westerhoff, H.V. (2007) Mathematical and theoretical biology for systems biology, and then ... vice versa. *J. Math. Biol.* **54**:147–150.

Westerhoff, H.V. & Kell, D.B. (1987) Matrix method for determining the steps most rate-limiting to metabolic fluxes in biotechnological processes. *Biotechnol. Bioeng.* **30**: 101–107.

Westerhoff, H.V. & Van Dam, K. (1987) *Thermodynamics and Control of Biological Free-Energy Transduction*. Elsevier, Amsterdam.

Westerhoff, H.V. & Van Workum, M. (1990) Control of DNA structure and gene expression. *Biomed. Biochim. Acta* **49**: 839–853.

Westerhoff, H.V., Koster, J.G., Van Workum, M. & Rudd, K.E. (1990) On the control of gene expression. In: *Control of Metabolic Processes* (ed. A. Cornish-Bowden). Plenum Press, New York, pp. 399–412.

5

Using mathematical models to probe dynamic expression data

Nick Monk

1 Introduction

Dynamical processes in cells and tissues can be understood as reflections of the dynamics of complex networks of interactions between sets of molecular components, operating both within and between cells (Zhu *et al.*, 2007). A central aim of systems biology is to gain mechanistic insight into the ways in which cellular and tissue-level function emerge from the underlying networks. The aim of this chapter is to describe methodologies that can be used for this purpose, and in particular to explore the role of mathematical modelling in extracting network information from dynamic expression data.

Within a network, the interactions between components are governed by basic biological mechanisms. The forms of these basic interactions are limited in number, and the wealth of complex behaviours exhibited by biological systems results primarily from the ways in which these mechanisms can be combined. In this context, 'basic biological mechanisms' refer to events involving the components of the system that directly effect changes in the state of the components. Examples include the creation and degradation of a component (which may or may not depend on interactions between the components). In cellular systems, the components are the constituent molecular species (mRNAs, proteins, metabolites, ions, etc.).

This chapter addresses the problem of how much can be determined about the basic biochemical mechanisms that underlie specific cellular dynamics. While specific examples of genetic regulatory networks are given, involving interactions between genes, mRNAs and proteins, the principles discussed are applicable more generally. The approaches used to answer this question depend on the type and quality of data available. The recent surge of interest in systems approaches to cell biology is in large part due to the emergence of new technologies that allow the state of large numbers of cellular components to be assayed simultaneously. To illustrate how these types of data can be used to gain quantitative insight into the structure and function of cellular networks, this chapter focuses on recent studies of three specific networks. These examples show how mathematical models provide a powerful tool for extracting information from data that is difficult to obtain by direct experimental manipulation of a system.

2 The structure of interaction networks

There are three different types of information required to characterize a cellular interaction network fully. First, there are the identities of the essential molecular components ('the parts list'). Second, there is the topology of the network (i.e. the list of all possible molecular interactions between network components). Third, there are the natures (or functional forms) of the interactions between the components. At its most basic level, functional form may be qualitative (e.g. an interaction may be 'activating' or 'repressive'). However, full characterization requires a more quantitative description of the way in which the strength of interaction depends on the states of the interacting components. Cellular networks typically have heterogeneous parts lists, and the spatial localization of each of these components within a cell may play an important role in the function of the network. Distinct spatial localizations of a molecular species can be represented in the network by multiple network components (for example, a transcription factor may be represented by two components reflecting its localization in nuclear or cytoplasmic compartments).

Until recently, acquisition of all the information required to furnish a detailed characterization of even small interaction networks was very time consuming and labour intensive. Notable examples of well-characterized networks include those underlying circadian rhythms (Gallego and Virshup, 2007), the yeast cell cycle (Chen *et al.*, 2004; Tyson *et al.*, 2002), bacterial chemotaxis (Bray *et al.*, 2007; Tindall *et al.*, 2008) and segmentation of the embryo of the fruit fly *Drosophila melanogaster* (Akam, 1987; Ingham, 1988). Recent technological advances, such as protein-interaction screens (Gavin *et al.*, 2002; Ito *et al.*, 2001; Uetz *et al.*, 2000), RNAi screens (Nybakken *et al.*, 2005) and transcription factor binding (ChIP) screens (Harbison *et al.*, 2004) have had a huge impact on the speed and ease with which network components and their (sometimes only putative) interactions can be acquired (Blais and Dynlacht, 2005).

While it is now a realistic (although still challenging) prospect to map out the interaction network underlying a wide range of cellular processes, the functional characterization of network interactions remains largely low-throughput and labour intensive. Furthermore, while qualitative insight into the nature of the interactions can be gained from standard genetic manipulations (e.g. mutations, mis-expression, knock-downs), quantitative information is much harder to obtain experimentally. A principal problem is the difficulty in achieving controlled, and possibly subtle, quantitative modulation of the states of the interacting components. For example, while it is relatively easy to control substrate or enzyme availability in an *in vitro* enzymatic reaction (and thus obtain a quantitative measure of the dependence of the reaction rate on reactant concentrations), this is very difficult to achieve for reactions within a cell.

This limitation on the accessibility of quantitative information on cellular interactions presents an important limitation on our ability to characterize interaction networks experimentally. Even simple networks with specified topology and qualitative forms of interaction can generate quite different dynamics, dependent on the quantitative details of key interactions. Typical examples include genetic 'toggle switches' (Cherry and Adler, 2000) and feed-forward motifs (Ingram *et al.*, 2006; Wall *et al.*, 2005). Thus, qualitative information alone is insufficient to explain qualitative features of the overall dynamics of a network (Lazebnik, 2002). Furthermore, while the observed dynamics may correspond (qualitatively) with one of the possible dynamic modes of the network, it is impossible to say how robust this behaviour might be, or how it might be manipulated.

3 Mathematical models of interaction networks

To move towards a more quantitative description of interaction networks, mathematical models are invaluable. Since the quantitative details of a network are important determinants of its system-level behaviour, it is important to be able to encode these details in a precise and logical manner. Mathematical models provide a natural framework for doing this, and are of two broad types. *Graphical models* encode the topology of the network as a formal graph, with each component represented by a vertex and each interaction by an edge in the graph (i.e. a link between two vertices). These models allow a range of quantitative measures of network structure to be determined (Albert, 2005; Barabási and Oltvai, 2004). *Dynamic models* supplement graphical models with specific functional representations of each interaction. These take the form of state evolution models, in which the state of the network $S(t)$ at a given time t represents the quantities of every network component, and network interactions are represented by a set of mathematical functions f that govern transitions of the state over time:

$$S(t_2) = f(S(t_1), t_2, t_1), t_2 > t_1. \tag{1}$$

The functions f may represent either deterministic or stochastic state evolution, resulting in a wide range of different modelling formalisms (de Jong, 2002).

To gain quantitative insight into a network, it is necessary to obtain quantitative data about the state of the network. If we consider the state to be represented by the quantities (and possibly sub-cellular localizations) of the network components, then these data are termed *quantitative expression data*. These take the form of time-course data: measurements of the quantity (or expression level) of individual network components at a series of time points. In this context, the aim of a systems approach is to gain insight into the basic mechanisms underlying a given system-level behaviour given a set of quantitative time-course expression data. The examples discussed in this chapter illustrate techniques for achieving this aim that take into account the quantity and quality of data that are available.

4 Linking state evolution models to expression data

For a network comprising n components (i.e. distinct molecular species) X_i, the network state $S(t)$ can be written as a vector comprising n real numbers $x_i(t)$ ($i = 1, 2, ..., n$) representing the quantities of each network component: $S(t) = \{x_1(t), x_2(t), ..., x_n(t)\}$. A set of time-course expression data takes the form $D = \{\xi_i(t_j), i = 1, 2, ..., n; j = 1, 2, ..., m\}$, where the $\xi_i(t_j)$ are measurements of the quantity of the network component X_i at a set of times t_j. Network inference aims to infer as much as possible about the forms of the functions f in Equation (1) using quantitative time-course expression data. Inference of biological networks presents particular problems because of the quality and quantities of time-course expression data that are available for biological processes (for a recent review, see Jaqaman and Danuser, 2006). Typical data sets present three principal challenges. First, data are usually available only for a subset of the network components (i.e. the data set is incomplete). Second, the number of time points in a data set is often rather small, with relatively large time intervals between successive measurements. These are issues that can, in principle, be addressed by increased effort in data acquisition. A third challenge is the high level of variability in measurements of each variable, which is a reflection of both measurement error and the 'noisy' cellular environment (Raser and O'Shea, 2005). Treatment of noise is beyond the scope of the current chapter.

At its most basic level, inference is a purely computational problem, and the role of the mathematical model is primarily to allow the network to be represented in an appropriate form that allows the goodness of fit between the model behaviour and the experimental data to be assessed quantitatively. At this level, it is not necessary to incorporate specific considerations about the basic biochemical mechanisms that underlie the changes in the variables over time; rather, the problem reduces to finding an appropriate time-dependent function that provides a good approximation to the time-course data. While computational approaches such as machine learning can often find mathematical functions that can closely approximate expression data, these do not necessarily yield significant insight into the basic biochemical mechanisms that play an important role in generating the observed behaviour of the system.

In order to go beyond purely computational approaches to inference, it is necessary to be more specific about the form of state evolution model that is used to represent the network. In particular, if we wish to gain insight into basic biochemical mechanisms, then these must be represented explicitly in the model. State evolution models that are based on representations of basic biochemical mechanisms operating at the level of the network components are described as *mechanistic models*. For concreteness, I shall consider one specific mathematical framework based on differential equations; however, the principles exemplified can be carried over to other types of models.

5 Differential equation models of cellular interaction networks

Differential equation (DE) models are sets of equations that specify the *instantaneous rate of change* of each variable as a function of the other variables in the model with which it interacts directly (i.e. with which it shares an edge in the graphical representation of the network). Mathematically, the general form of a (non-delayed) differential equation model for an n-component network is:

$$\frac{dx_i}{dt}(t) = f_i\left(\left\{\langle x_i \rangle_j (t), j = 1, 2, \ldots, N_i\right\}, \boldsymbol{\pi}_i\right), \quad i = 1, 2, \ldots, n, \tag{2}$$

where $\frac{dx_i}{dt}(t)$ denotes the instantaneous rate of change of variable x_i at time t, $\left\{\langle x_i \rangle_j (t), j = 1, 2, \ldots, N_i\right\}$ is the set of N_i variables that interact directly with x_i, and the f_i are mathematical functions that represent the basic mechanisms that contribute to changes in the variable x_i (dependent on a set of parameters $\boldsymbol{\pi}_i$).

There are a number of implicit assumptions that are encoded in this type of model. For example, the spatial location of the network components is not modelled, and it is assumed that the network components are 'well mixed' in an essentially spatially homogeneous cell (for an example of spatially-explicit differential equation models see Clayton and Nash, 2008). In addition, although the quantities (numbers) of each component must be positive integers, they are approximated as continuously varying real numbers, which can change by arbitrarily small amounts over time. In cases where these assumptions do not provide a reasonable approximation, other mathematical formalisms (such as stochastic models) are more appropriate (see, for example, Bower and Bolouri, 2001). A further assumption that is encoded in the form of Equation (2) is that the interactions have instantaneous effects. In reality, certain cellular interactions, such as transcription and translation, involve significant time delays (see, for example, MacDonald, 1989; Monk, 2003).

In order to use a DE model for inference it is first necessary to calculate the time-courses predicted for each network variable by Equation (2). Typically, this must be done computationally, for a given set of functions f_i, parameters π_i and starting values (at some initial time) for each of the variables (for an excellent practical guide, see Shampine et al., 2003). However, a valuable feature of DE models is the existence of a powerful mathematical 'toolkit' (the theory of dynamical systems) that allows a range of quantitative and qualitative aspects of the time-courses to be deduced from the form of the differential equations without determining the time-courses explicitly (see, for example, Jordan and Smith, 1999). This provides a means of going beyond inference in cases where only limited experimental data are available, to explore the relative roles of different combinations of interactions (see below).

The principal issue that must be addressed in developing a DE model of a network is the choice of the functions f_i (and the associated parameters π_i). For a typical cellular process it is reasonable to assume that a small set of basic molecular mechanisms contribute to interactions between network components that result in changes in network state. Examples include transcription (production of mRNA regulated by specific transcription factors), translation (production of protein from mRNA templates), degradation (e.g. proteolysis) and enzymatic conversions (e.g. protein phosphorylation). If it is assumed that each of these mechanisms operates essentially independently of the others (so that, for example, the rate of transcription at a given time does not depend explicitly on the rate of translation), then the functions f_i can be built up piecewise from mathematical representations of these basic processes.

To make this approach explicit, consider a simple representation of an autoregulatory transcription factor (i.e. a transcription factor that regulates the rate of transcription of its own gene directly). Let the state of this system be represented by the quantities of the corresponding mRNA and protein, denoted by $m(t)$ and $p(t)$, respectively. Assuming that the dynamics of this system are governed primarily by four basic biochemical processes (transcription, translation, mRNA degradation and protein degradation), then a DE model can be written as:

$$\frac{dm}{dt} = \left(\text{transcription rate}\right) - \left(\text{mRNA degradation rate}\right)$$

$$\frac{dp}{dt} = \left(\text{translation rate}\right) - \left(\text{protein degradation rate}\right).$$

The first step in specifying functional forms for each basic biochemical process is the assignment of the functional dependencies of each reaction. So, for example, it is reasonable to assume that the transcription and translation rates depend only on the quantities of protein and mRNA, respectively, while the degradation rates depend only on the quantities of the component being degraded. Making the further assumption that the degradation rate of each component is proportional to the quantity of that component (linear degradation) yields the following model:

$$\frac{dm}{dt} = f_M(p) - d_1 m$$
$$\frac{dp}{dt} = f_P(m) - d_2 p,$$

(3)

where f_M and f_P are functions representing transcription and translation respectively, and d_1 and d_2 are parameters that specify the degradation rates of mRNA and protein respectively. Equation (3) provides a simple framework for building models centred on transcriptional regulation, which can be supplemented with additional interaction terms (such as mass-action representations of protein–protein interactions involving transcription factors). Specific information relating to the structure of the regulatory DNA sequences controlling the rate of transcription of each component in the network can be encoded in the mathematical forms of the transcription functions. If a network contains n transcription factors p_i (and corresponding mRNAs m_i) and the transcription rate of m_i is regulated by the set of proteins $\langle \mathbf{p} \rangle_i$, then the model can be written as:

$$\frac{dm_i}{dt} = f_{Mi}(\langle \mathbf{p} \rangle_i) - d_{1i} m_i$$
$$\frac{dp_i}{dt} = f_{Pi}(m_i) - d_{2i} p_i, \quad i = 1, 2, \ldots, n.$$

(4)

This standard form of model has been used successfully to capture the essential features of many different cellular networks whose dynamics are governed primarily by transcriptional regulation (see, for example, Fall *et al.*, 2002; Tyson *et al.*, 2003).

6 Extracting information from expression data

Given the specific context of quantitative time-course expression data and a suitable mathematical model, the aim of gaining mechanistic insight into the ways in which cellular and tissue-level function emerge from the underlying networks can be expressed in a more concrete way. Time-course expression data contain information about the nature of the basic interactions occurring between network components, albeit in a highly convoluted form. The aim of a systems approach can be seen as to extract useful (and reliable) information about these interactions. In this sense, the problem is essentially one of bioinformatics – time-course expression data are simply high-dimensional datasets from which we wish to extract information on the underlying system. Mathematical models (such as Equation (4)) are tools that can be used to probe these data, in order to extract specific information.

The way in which mathematical models can be used as tools to extract information from time-course expression data depends on both the quality and quantity of the expression data, and on the extent to which the important network components and topology are known. In the following, three examples that illustrate different scenarios are outlined. These are by no means exhaustive; rather, they are intended to illustrate a number of key points.

6.1 *Example 1: Inference of a transcription factor expression profile, given time-course mRNA expression data for its target genes*

It is often the case that it is difficult to obtain quantitative time-course expression data for a particular class of network components. For example, while microarray technology provides a convenient and well-established means of assaying the expression of thousands of mRNAs simultaneously (Baldi and Hatfield, 2002), it is typically more challenging to assay quantitative expression levels for multiple proteins (Albeck *et al.*, 2006). Because of this practical constraint, it is much more common to find time-course data for mRNA expression than for protein expression. However, it is clearly no less important to understand the dynamics of protein expression than that of mRNA expression. How can mathematical models be used to help infer protein profiles from time-course mRNA expression data?

To see how this can be achieved, consider the simple task of deducing the expression profile of a single transcription factor from time-course expression data for its direct transcriptional targets (Barenco *et al.*, 2006; Lawrence *et al.*, 2007). In this case, there is a single transcription factor with expression level $p(t)$, regulating the transcription rate of a number of target mRNAs $m_i(t)$, $i = 1, 2, ..., n$. Using the approach outlined above, the expression of the target mRNAs can be modelled as follows:

$$\frac{dm_i}{dt} = f_i\big(p(t)\big) - d_i m_i, \quad i = 1, 2, ..., n, \tag{5}$$

where the functions f_i encode information about how the transcription factor regulates the transcription rate of each target, and the d_i are the linear degradation rates of the targets. Given the model encoded by Equations (5) and a set of time-course expression data for the targets, can anything be deduced about the form of the unknown transcription factor profile $p(t)$? Barenco *et al.* (2006) address this question for the tumour suppressor transcription factor p53 by assuming a linear form for the transcription functions – that is, that the rate of transcription of each mRNA increases in proportion to the amount of transcription factor:

$$\frac{dm_i}{dt} = a_i p(t) + b_i - d_i m_i, \quad i = 1, 2, ..., n, \tag{6}$$

for some set of positive parameters a_i and b_i. The a_i provide a measure of the sensitivity of each target to changes in the transcription factor and the b_i are the basal transcription rates in the absence of any transcription factor.

Using this model, Barenco *et al.* develop computational strategies for inferring both the unknown model parameters (the values a_i, b_i and d_i) and the unknown transcription factor profile $p(t)$ for a small set of known targets for which time-course microarray data are available. Barenco *et al.* find that unconstrained inference (i.e. allowing the parameters to take any positive values) results in estimates for the parameters and for $p(t)$ that have very high variance. In other words, in the absence of any prior knowledge of the model parameters, it is not possible even for this simple system to infer with any confidence the likely values of either the model parameters or the input function $p(t)$ by using the available mRNA expression data. This supports previous results obtained for hypothetical network structures by Zak *et al.* (2003).

The failure of the unconstrained inference scheme employed by Barenco *et al.* is a reflection of the fact that there are too many free parameters in the model as posed. This problem can be overcome by imposing constraints on the allowed values of at least one of the model parameters. Barenco *et al.* constrain the model by measuring the degradation rate of just one target mRNA, and show that this results in a dramatic reduction in the variance of the estimates for the remaining model parameters and the input function. Again, this result is in concordance with the results of Zak *et al.* (2003), who showed that successful inference of a class of hypothetical transcription networks depends on the prior specification of mRNA degradation rates. The findings of Barenco *et al.* have been reproduced using the same data but a different inference method by Lawrence *et al.* (2007). Lawrence *et al.* also consider non-linear transcription rate functions, encoding the essential non-linearity of typical transcriptional responses, and show that model parameters can also be inferred successfully in this case.

The results of Barenco *et al.* show how quantitative time-course data for a subset of network components can be used to infer both the parameters representing basic biochemical properties of these components (such as mRNA stability) and the expression profile of additional network components for which explicit time-course data are not available. The ability to extract this information from the expression data requires only the specification (from separate measurements) of a single mRNA degradation rate. Importantly, Barenco *et al.* validated their predictions by experimentally manipulating p53 levels. In addition to providing valuable information on quantitative aspects of this simple transcriptional network, the inferred expression profile for p53 can be used to probe genomic-scale mRNA expression data for additional putative targets of p53 (note that only a small number of known p53 targets were used for inference). Barenco *et al.* use their model to generate a ranked list of putative p53 targets and show that this method is more accurate than the more conventional approach of clustering based on mRNA time-course data.

This example illustrates the power of mathematical models in inferring unknown expression profiles from partial network data, and how the information gained can be used in focusing further experimental studies aimed at identifying additional network components. It is important to note that this approach has so far only been shown to be successful for rather simple network structures (in this case, the response of multiple targets to a single transcription factor). How successful (and how computationally costly) this approach would be when applied to more complex network structures remains to be seen.

6.2 *Example 2: Inference of the regulatory structure of a transcription network using protein expression data*

In the previous example, expression data were only available for a subset of network components, but the network topology was known with confidence (each target gene had a single input – p53). However, it is typically the case that the topology of more complex networks is difficult to determine with confidence from experimental data. This is particularly so when the transcription rate of a network component depends on the combinatorial action of multiple transcription factors. In the case of uncertain network topology, mathematical models can be used to probe time-course expression data to infer not only network parameters but also network topology itself. Clearly, successful inference of the topology of the entire network depends on the availability of expression data for every network component (for a discussion of partial network inference, see Perkins *et al.*, 2004).

An example of a well-characterized network for which complete high quality time-course expression data are available is the gap gene network that operates during the early stages of segmentation of the embryo of the fruit fly *Drosophila*. The gap genes encode a highly-connected network of transcription factors that are expressed in broad dynamic domains along the anterior-posterior axis of the *Drosophila* embryo. The network operates before the cellularization of the embryo, during the syncytial blastoderm stage, when the embryo consists of individual nuclei embedded in a shared cytoplasm. Interpretation of genetic analyses of the segmentation process allow a more-or-less complete parts list of the gene/protein interaction network operating in the blastoderm, and gives a qualitative picture of key network interactions (Akam, 1987; Ingham, 1988; Nüsslein-Volhard and Wieschaus, 1980).

In order to gain a detailed quantitative picture of the dynamics of the gap gene network, John Reinitz and colleagues undertook the task of determining the time-course expression profile of each gap gene product in nuclei along the entire axis of the embryo (Janssens *et al.*, 2005; Kosman *et al.*, 1998; Poustelnikova *et al.*, 2004). The resulting data provide a complete set of quantitative time-course profiles for the entire gap network, with a time resolution of around 6 minutes. In contrast to the simple network structure assumed for the exploration of the p53 target data in the previous example, genetic data suggest that the gap gene network has a more complex topology, with each gene being regulated combinatorially by multiple gap gene products. Furthermore, the detailed expression data reveal subtle features of the dynamics of the gap gene products that are difficult to account for using only qualitative descriptions of the interactions between network components. The availability of these data therefore raises the question of how much quantitative insight into the operation of the gap gene network can be inferred from the data using a mathematical model.

Unlike the p53 network considered in the previous example, the gap network operates in a spatially explicit context. However, by representing the expression level of each gap gene product in each nucleus as a separate model variable, the spatial network can be modelled using a system of coupled differential equations of the type in Equation (4). This mathematical representation is appropriate since the gap proteins are localized predominantly in the nuclei, so that their expression pattern is essentially spatially discrete. Jaeger *et al.* (2004a, 2004b) use a model of this type to explore the information contained in the gap gene time-course expression data. Rather than assuming a given network topology, the model is based on a fully connected network in which all possible pairwise interactions between network components are allowed (Mjolsness *et al.*, 1991). In this approach, each pairwise interaction corresponding to transcriptional regulation is represented in the model by an interaction weight parameter.

By minimizing the discrepancy between the spatio-temporal expression profiles predicted by the model and the observed profiles, Jaeger *et al.* infer optimal values for all the model parameters, including the interaction weights (Jaeger *et al.*, 2004a). Since this model-based approach allows the network interaction weights to be inferred from the data, it consequently allows the quantitative topology of the network to be inferred from the data (since the topology is encoded in the set of weights). This example therefore demonstrates that if high quality time-course expression data are available for all network components, then it is possible to use a mathematical model to infer the detailed quantitative topology of the network. Of course, this approach requires the specification of mathematical functions representing transcriptional regulation. At present, there is no single agreed way of representing the combinatorial regulation of transcription by multiple transcription factors, and Jaeger *et al.* use a single sigmoidal function for each gap gene product, with weighted inputs from all gap gene products. The robustness of the inferred network topology to changes in the detailed form of these functions has since been confirmed by using a range of different mathematical representations of the transcription functions (Perkins *et al.*, 2006). To reduce computational cost, the models used by Jaeger *et al.* and Perkins *et al.* consider only the protein products of the gap genes, with the corresponding mRNAs remaining implicit. While this model simplification impacts on the specific parameter values that can be inferred from the data, it is unlikely to have a significant effect on the inferred network toplogy.

The ability to infer a detailed quantitative network topology from expression data opens up the possibility of using the resulting network model as a proxy for the real network. One motivation for doing this stems from the central aim of a systems approach to cellular networks stated at the start of this chapter – to gain mechanistic insight into the ways in which cellular and tissue-level dynamics emerge from underlying interaction networks. In the context of the *Drosophila* blastoderm, one aspect of this aim is to understand the quantitative role played by individual network interactions in determining the dynamics of the observed spatio-temporal expression of the gap gene products. This type of detailed analysis cannot be achieved using experimental manipulations alone, because of the impossibility of isolating and modulating each interaction in turn. In contrast, it is straightforward to do this in the model. By using the inferred model as a proxy for the real gap gene network, Jaeger *et al.* perform a detailed analysis that reveals the relative contributions of each individual network interaction to different aspects of the overall expression patterns (Jaeger *et al.*, 2004a, 2004b). As with all insights gained from models, the necessary confirmation of their validity can only come from detailed experiments. However, the combination of high quality data acquisition and data-driven modelling used by Reinitz and colleagues provides a powerful example of how a detailed quantitative understanding can be gained for complex spatio-temporal cellular processes.

7 Using mathematical models to explore partial networks

The analysis of the gap gene network outlined above shows how a complete set of time-course data can be used to perform a detailed dissection of a biological process for which the full set of relevant network components is known with confidence. However, there are few processes for which the essential network components are known with such confidence. What use can be made of mathematical models in probing time-course expression data for such processes? As with the examples above, a model based on the known network components can be constructed, and used to infer parameters from available expression data. However, given the lack of confidence in the completeness of the model, the confidence in parameter inference will be correspondingly low. In such cases, parameter validation can provide valuable constraints on the model (as was seen in Barenco *et al.*, 2006) and a resulting increase in the confidence of inferred parameters.

The potential importance of parameter validation for partial networks raises the question of whether the parameter(s) to be determined experimentally can be chosen rationally. While practical considerations typically play a major role in this choice (for example, degradation rates are typically much easier to measure than the affinities of transcription factors for their binding sites on DNA), can insight gained from mathematical models be used to focus directed functional studies of network interactions? This question becomes increasingly relevant as the size of the network increases, since the number of model parameters grows more quickly than the number of components, and it rapidly becomes impractical for all parameters to be determined.

One approach to the problem of selection of parameters for detailed experimental investigation relies on an assessment of the degree of the sensitivity of key quantitative features of the model dynamics on the model parameters (Ihekwaba *et al.*, 2004; Ingalls and Sauro, 2003. Assuming that it is possible to infer an appropriate set of model parameters from available expression data (i.e. a set for which the behaviour of the model matches the data closely), the aim of a *parameter sensitivity analysis* is to determine how quickly the goodness of fit of the model behaviour decreases as each model parameter is changed from its inferred value (keeping the values of the remaining parameters fixed). Key model parameters – those for which the model dynamics are very sensitive to parameter changes – provide prime targets for detailed experimental investigation.

A potenial limitation of parameter sensitivity analysis is that it only considers the effects of variations in one parameter at a time. This can present problems when the expression data cannot be interpreted directly in terms of absolute numbers (or concentrations) of network components. If this is the case (as it is for most expression data), then fitting a model to the data cannot determine the absolute scale of expression, leaving an element of uncertainty in parameter inference. When fitting to the 'form' of expression data (rather than the absolute level), it is often the case that specific *combinations* of parameters determine network behaviour (modulo overall scaling), and it is these combinations that should be used for inference.

A simple example of how mathematical models can be used to determine critical parameter combinations is provided by a model of an autoregulatory loop centred on the murine transcription factor Hes1. The Hes family of transcriptional repressors are involved in a wide range of patterning events during vertebrate development

(Kageyama *et al.*, 2007). The Hes1 protein, and its corresponding mRNA, exhibit striking oscillatory expression with a period of around two hours in a number of different cell types during development, and these oscillations are believed to regulate the timing of other events in these cells (Kageyama *et al.*, 2007). In the context of the general discussion above, these oscillations represent a simple example of a system-level behaviour with a number of important features such as the period and amplitude of the oscillations. If Hes1 oscillations are controlling developmental timing, then it is important to understand how the oscillatory period depends on the basic biochemical mechanisms operating in the network of interactions involving Hes1.

Hes1 is a basic helix-loop-helix transcription factor that negatively regulates its own expression by binding directly (in dimeric form) to at least four binding sites in the regulatory sequence of the *hes1* gene (Takebayashi *et al.*, 1994). The mechanistic origin of Hes1 oscillations is difficult to address directly in the context of the mouse embryo, where Hes1-dependent processes involve numerous additional factors, and where it is difficult to obtain well-resolved time-course expression data for Hes1. Fortunately, Hes1 oscillations can be observed in a much simpler system, since oscillatory Hes1 expression can be induced by serum stimulation of a range of murine cell types in culture (Hirata *et al.*, 2002). Importantly, in cells such as fibroblasts, serum stimulation does not induce expression of many of the network components that interact with Hes1 in the embryo, increasing the likelihood that the oscillations are the result of a network of interactions centred on Hes1 itself. Furthermore, in a population of induced cells the oscillations are at least initially in phase, making it possible to detect oscillations in mRNA and protein expression levels in samples drawn from a population at regular time intervals (Hirata *et al.*, 2002).

Assuming that Hes1 oscillations depend primarily on the negative autoregulation of Hes1, a simple model can be written in terms of two variables $M(t)$ and $P(t)$, representing the quantities of mRNA and protein in a cell, respectively (Monk, 2003):

$$\frac{dM}{dt} = k_M \frac{P_0^b}{P_0^b + P(t-\tau_1)^b} - d_M M(t)$$

$$\frac{dP}{dt} = k_p M(t-\tau_2) - d_p P(t). \tag{7}$$

In this model, specific forms for the functions representing the rates of transcription and translation have been chosen. The meanings of the parameters are as follows: k_M is the maximal rate of *hes1* transcription (in the absence of Hes1 protein); k_p is the rate at which Hes1 protein is produced by translation from a single *hes1* mRNA molecule; d_M and d_p are the linear degradation rates of *hes1* mRNA and Hes1 protein, respectively; τ_1 and τ_2 are time delays representing the delays inherent in the transcription and translation processes, respectively; P_0 is the level of Hes1 protein for which *hes1* transcription is reduced to half its maximal value (the repression threshold); h provides a measure of the sensitivity of the transcription rate to changes in the level of Hes1 protein. *Figure 1* illustrates the way in which the transcription function depends on P_0 and h.

This model, simple as it is, has eight parameters. By simple manipulation of the model, bearing in mind that the time-course data obtained by Hirata *et al.* (2002) do

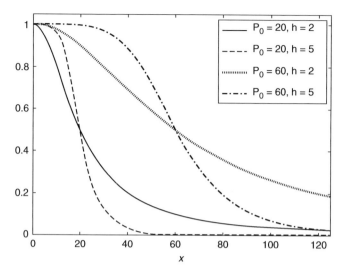

Figure 1. *Illustration of the dependence of the form of the transcription rate function* $f_M(x) = P_0^h / (P_0^h + x^h)$ *on the parameters* P_0 *(the repression threshold) and* h *(the sensitivity).*

not provide information on the absolute levels of mRNA and protein expression, it is possible to show that the nature of the oscillatory behaviour of the model actually depends on only six independent parameters. Furthermore, the form of the expression profiles of mRNA and protein, considered separately, depend on only five parameters (the sixth affects only the relative timing of the two profiles). Since the absolute expression levels cannot be determined from data, the mRNA and protein variables can be rescaled:

$$m(t) = \frac{M(t)}{k_M}, \quad p(t) = \frac{P(t)}{k_M k_P},$$

yielding the following form of the model equations:

$$\frac{dm}{dt} = \frac{p_0^b}{p_0^b + p(t - \tau_1)^b} - d_M m(t)$$

$$\frac{dp}{dt} = m(t - \tau_2) - d_p p(t),$$

(8)

where $p_0 = P_0/(k_M k_P)$ is the scaled repression threshold. This simple rescaling shows immediately that the three parameters P_0, k_M and k_P do not act independently, but rather in the specific combination $p_0 = P_0/(k_M k_P)$. It is this effective parameter that determines the form of the mRNA and protein expression profiles (modulo unknown absolute scales). It would thus make little sense to attempt to use the available expression data to infer values for these parameters independently. Furthermore, by introducing a time-shifted mRNA variable $x(t) = m(t - \tau_1)$, it can be seen that

$$\frac{dx}{dt} = \frac{p_0^b}{p_0^b + p(t-\tau)^b} - d_M x(t)$$

$$\frac{dp}{dt} = x(t) - d_p p(t),$$

(9)

where $\tau = \tau_1 + \tau_2$. The only effect of this time shift is to change the relative timing of the mRNA and protein profiles by an amount τ_2. Since the available expression data give little information on these relative timings, the model encoded in Equation (9) is more appropriate for parameter inference than the form in Equation (7).

In addition to determining time-course data for the full Hes1 system, Hirata *et al.* (2002) also assay the time-course of mRNA and protein populations in the absence of transcription and translation, respectively. This allows the direct inference of values for the parameters d_M and d_p (both are inferred to be approximately 0.03 min^{-1}). Thus, adopting these values in the full model (9) leaves only three parameters to be inferred from the data.

However, it is possible to get further insight into the system before embarking on inference. Consideration of the form of the model (9) shows that sustained oscillations can be generated only if the total delay τ exceeds a certain critical value, which depends on the repression threshold and sensitivity (Monk, 2003). Furthermore, for the sustained oscillations to have the observed period of 2 hours (Hirata *et al.*, 2002; Masamizu *et al.*, 2006), the sensitivity h must be at least 5. Finally, it can be shown both analytically and by numerical simulation that the period of oscillation of the system depends very weakly on the value of the repression threshold p_0 over a wide range of biologically plausible values (see *Figure 2*).

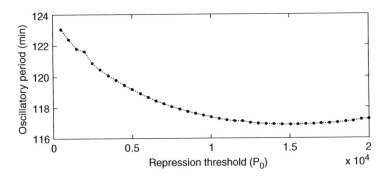

Figure 2. *Relative insensitivity of the period of oscillation of hes1 mRNA and Hes1 protein on the repression threshold. The period (determined by numerical simulation of Equation (7)) varies by only 5% around its mean value of 120 min for values of P_0 ranging from 500 to 20 000 molecules. Numerical simulation performed using the following parameter values: $k_M = 33$ min^{-1}, $k_p = 4.5$ min^{-1} per mRNA molecule, $d_M = 0.03$ min^{-1}, $d_p = 0.03$ min^{-1}, $\tau_1 = 16.2$ min, $\tau_2 = 2$ min, $h = 5$.*

Using the insight gained from consideration of the general properties of the model used to represent the Hes1 system, and inferred values of the two degradation rates, the inference problem can be reduced to one depending on only two parameters: h and τ.

Even using the insight gained from analysis of the model equations to reduce the problem to that of inferring h and τ from the time-course data, it is not possible to achieve a good fit to the data without assigning infeasibly high values to the sensitivity parameter h (which provides a measure of the degree of cooperativity between the Hes1 binding sites in the *hes1* gene). The main reason for this failure is the inability of the model described by Equation (7) to generate sustained oscillations with a large amplitude (defined as the fold-difference between minimum and maximum expression levels). Time-course data show that both *hes1* mRNA and Hes1 protein oscillations exhibit fold-difference amplitudes of around 3–8 (Hirata *et al.*, 2002), and one would expect that a sufficiently large oscillatory amplitude would be necessary for Hes1 oscillations to be physiologically relevant. The model described by Equation (7) is incapable of generating protein oscillations with suitable high amplitude (see *Figure 3*), suggesting that the model does not capture the true nature of the network underlying Hes oscillations.

This failure could result from two possible shortcomings of the model. First, it could be that the network contains components other than *hes1* mRNA and Hes1 protein that are important in generating high-amplitude oscillations. Alternatively, it might be that the two components are actually capable of generating higher-amplitude oscillations, but that the level of mechanistic detail in the current model is insufficient. While the former possibility could well be true, existing extensions of the model do not appear to address the amplitude problem (Bernard *et al.*, 2006).

The second possibility can be explored through the analysis of a more detailed mechanistic model of the Hes1 network, incorporating explicit representations of basic biochemical mechanisms known to be involved, such as Hes1 dimerization and nucleo-cytoplasmic shuttling. With this more detailed mechanistic model, it does become possible to find biologically reasonable parameters that result in the generation of oscillations with an amplitude in agreement with the time-course data (N. Monk and H. Momiji, unpublished data). Interestingly, inference using the more detailed

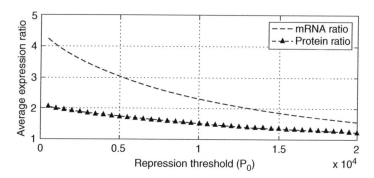

Figure 3. *Dependence of oscillatory amplitude of hes1 mRNA and Hes1 protein expression on the repression threshold (P_0). Other parameters as in Figure 2.*

model predicts that the monomeric form of the Hes1 protein should be significantly less stable than the dimeric form, a phenomenon that has been observed for a number of bacterial transcription factors (Buchler *et al.*, 2005).

8 Concluding remarks

The increasing availability of quantitative time-course data for the expression of the molecular constituents of cellular networks provides the genuine prospect of inferring detailed mechanistic pictures of the workings of complex biological networks. If a reliable 'parts list' of a network is available, mathematical models can be used directly to extract quantitative information about the network interactions, enabling a detailed dissection of the corresponding biological process. In the more common case where knowledge about the network is less reliable, models and data can be combined in an integrative approach that can help to identify the most promising areas of further experimental research (see, for example, Locke *et al.*, 2005).

References

Akam, M. (1987) The molecular basis for metameric pattern in the *Drosophila* embryo. *Development* 101: 1–22.

Albeck, J.G., MacBeath, G., White, F.M., Sorger, P.K., Lauffenburger, D.A. & Gaudet, S. (2006) Collecting and organizing systematic sets of protein data. *Mol. Cell Biol.* 7: 803–812.

Albert, R. (2005) Scale-free networks in cell biology. *J. Cell Sci.* 118: 4947–4957.

Baldi, P. & Hatfield, G.W. (2002) *DNA Microarrays and Gene Expression*. Cambridge University Press, Cambridge.

Barabási, A.-L. & Oltvai, Z.N. (2004) Network biology: Understanding the cell's functional organization. *Nature Rev. Genet.* 5: 101–113.

Barenco, M., Tomescu, D., Brewer, D., Callard, R., Stark, J. & Hubank, M. (2006) Ranked prediction of p53 targets using hidden variable dynamic modeling. *Genome Biol.* 7: R25.

Bernard, S., Cajavec, B., Pujo-Menjouet, L., Mackey, M.C. & Herzel, H. (2006) Modelling transcriptional feedback loops: the role of Gro/TLE1 in Hes1 oscillations. *Phil. Trans. R. Soc. A* 364: 1155–1170.

Blais, A. & Dynlacht, B.D. (2005) Constructing transcriptional regulatory networks. *Genes Dev.* 19:1499–1511.

Bower, J.M. & Bolouri, H. (2001) *Computational Modelling of Genetic and Biochemical Networks*. MIT Press. Cambridge, MA.

Bray, D., Levin, M.D. & Lipkow, K. (2007) The chemotactic behaviour of computer-based surrogate bacteria. *Curr. Biol.* 17: 12–19.

Buchler, N.E., Gerland, U. & Hwa, T. (2005) Nonlinear protein degradation and the function of genetic circuits. *Proc. Natl Acad. Sci. USA* 102: 9559–9564.

Chen, K., Calzone, L., Csikasz-Nagy, A., Cross, F., Novak, B. & Tyson, J. (2004) Integrative analysis of cell cycle control in budding yeast. *Mol. Biol. Cell* 15: 3841–3862.

Cherry, J.L. & Adler, F.R. (2000) How to make a biological switch. *J. Theor. Biol.* 203: 117–133.

Clayton, R. & Nash. M. (2008) Modelling the mammalian heart. Chapter 9, this publication.

de Jong, H. (2002). Modeling and simulation of genetic regulatory systems: A literature review. *J. Comput. Biol.* **9:** 67–103.

Fall, C.P., Marland, E.S., Wagner, J.M. & Tyson J.J. (2002) *Computational Cell Biology.* Springer, New York.

Gallego, M. & Virshup, D.M. (2007) Post-translational modifications regulate the ticking of the circadian clock. *Nat. Rev. Mol. Cell. Biol.* **8:** 139–148.

Gavin, A.-C., Bösche, M., Krause, R. *et al.* (2002) Functional organization of the yeast proteome by systematic analysis of protein complexes. *Nature* **415:** 141–147.

Harbison, C.T., Gordon, D.B., Lee, T.I. *et al.* (2004). Transcriptional regulatory code of a eukaryotic genome. *Nature* **431:** 99–104.

Hirata, H., Yoshiura, S., Ohtsuka, T., Bessho, Y., Harada, T., Yoshikawa, K. & Kageyama, R. (2002) Oscillatory expression of the bHLH factor Hes1 regulated by a negative feedback loop. *Science* **298:** 840–843.

Ihekwaba, A.E.C., Broomhead, D.S., Grimley, R.L., Benson, N. & Kell, D.B. (2004) Sensitivity analysis of parameters controlling oscillatory signalling in the NF-κB pathway: the roles of IKK and I-κBα. *Syst. Biol.* **1:** 93–103.

Ingalls, B.P. & Sauro, H.M. (2003) Sensitivity analysis of stoichiometric networks: an extension of metabolic control analysis to non-steady state trajectories. *J. Theor. Biol.* **222:** 23–36.

Ingham, P.W. (1988) The molecular genetics of embryonic pattern formation in *Drosophila. Nature* **335:** 25–34.

Ingram, P., Stumpf, M. & Stark, J. (2006) Network motifs: structure does not determine function. *BMC Genomics* **7:** 108.

Ito, T., Chiba, T., Ozawa, R., Yoshida, M., Hattori, M. & Sakaki, Y. (2001) A comprehensive two-hybrid analysis to explore the yeast protein interactome. *Proc. Natl Acad. Sci. USA* **98:** 4569–4574.

Jaeger, J., Surkova, S., Blagov, M. *et al.* (2004a) Dynamic control of positional information in the early *Drosophila* embryo. *Nature* **430:** 368–371.

Jaeger, J., Blagov, M., Kosman, D. *et al.* (2004b) Dynamical analysis of regulatory interactions in the gap gene system of *Drosophila melanogaster. Genetics* **167:** 1721–1737.

Janssens, H., Kosman, D., Vanario-Alonso, C.E., Jaeger, J., Samsonova, M. & Reinitz, J. (2005) A high-throughput method for quantifying gene expression data from early *Drosophila* embryos. *Dev. Genes Evol.* **215:** 374–381.

Jaqaman, K. & Danuser, G. (2006) Linking data to models: data regression. *Nat. Rev. Mol. Cell Biol.* **7:** 813–819.

Jordan, D. & Smith, P. (1999) *Nonlinear Ordinary Differential Equations: An Introduction to Dynamical Systems.* Oxford University Press, Oxford.

Kageyama, R., Ohtsuka, T. & Kobayashi, T. (2007) The Hes gene family: repressors and oscillators that orchestrate embryogenesis. *Development* **134:** 1243–1251.

Kosman, D., Small, S. & Reinitz, J. (1998) Rapid preparation of a panel of polyclonal antibodies to *Drosophila* segmentation proteins. *Dev. Genes Evol.* **208:** 290–294.

Lawrence, N.D., Sanguinetti, G. & Rattray, M. (2007) Modelling transcriptional regulation using Gaussian processes. In: *Advances in Neural Information Processing Systems* (eds B. Schölkopf, J.C. Platt and T. Hoffman). MIT Press, Cambridge, MA.

Lazebnik, Y. (2002). Can a biologist fix a radio? — Or, what I learned while studying apoptosis. *Cancer Cell* **2**: 179–182.

Locke, J.C.W., Southern, M.M., Kozma-Bognár, L. Hibberd, V., Brown, P.E., Turner, M.S. & Millar, A.J. (2005) Extension of a genetic network model by iterative experimentation and mathematical analysis. *Mol. Sys. Biol.* **1**: 2005.0013 (doi:10.1038/msb4100018).

MacDonald, N. (1989). *Biological delay systems: linear stability theory*. Cambridge University Press, Cambridge.

Masamizu, Y., Ohtsuka, T., Takashima, Y., Nagahara, H., Takenaka, Y., Yoshikawa, K., Okamura, H. & Kageyama, R. (2006) Real-time imaging of the somite segmentation clock: Revelation of unstable oscillators in the individual presomitic mesoderm cells. *Proc. Natl Acad. Sci. USA* **103**: 1313–1318.

Mjolsness, E., Sharp, D.H. & Reinitz, J. (1991) A connectionist model of development. *J. Theor. Biol.* **152**: 429–453.

Monk, N.A.M. (2003) Oscillatory expression of Hes1, p53 and NF-κB driven by transcriptional time delays. *Curr. Biol.* **13**: 1409–1413.

Nüsslein-Volhard, C. & Wieschaus, E. (1980) Mutations affecting segment number and polarity in *Drosophila*. *Nature* **287**: 795–801.

Nybakken, K., Vokes, S.A., Lin, T.-Y., McMahon, A.P. & Perrimon, N. (2005) A genome wide RNAi interference screen in *Drosophila melanogaster* cells for new components of the Hh signalling pathway. *Nature Genet.* **37**: 1323–1332.

Perkins, T.J., Hallett, M. & Glass, L. (2004) Inferring models of gene expression dynamics. *J. Theor. Biol.* **230**: 289–299.

Perkins, T.J., Jaeger, J., Reinitz, J. & Glass, L. (2006) Reverse engineering the gap gene network of *Drosophila melanogaster*. *PLoS Comp. Bio.* **2**: e51.

Poustelnikova, E., Pisarev, A., Blagov, M., Samsonova, M. & Reinitz, J. (2004) A database for management of gene expression data *in situ*. *Bioinformatics* **20**: 2212–2221.

Raser, J.M. & O'Shea, E.K. (2005) Noise in gene expression: origins, consequences, and control. *Science* **309**: 2010–2013.

Shampine, L.F., Gladwell, I. & Thompson, S. (2003) *Solving ODEs with Matlab*. Cambridge University Press, Cambridge.

Takebayashi, K., Sasai, Y., Sakai, Y., Watanabe, T., Nakanishi, S. & Kageyama, R. (1994) Structure, chromosomal locus, and promoter analysis of the gene encoding the mouse helix-loop-helix factor Hes-1. *J. Biol. Chem.* **269**: 5150–5156.

Tindall, M.J., Maini, P.K., Armitage, J.P., Singleton, C. & Mason, A. (2008) Intracellular signalling during bacterial chemotaxis. Chapter 8, this publication.

Tyson, J.J., Csikasz-Nagy, A. & Novak, B. (2002) The dynamics of cell cycle regulation. *Bioessays* **24**: 1095–1109.

Tyson, J.J., Chen, K.C. & Novak, B. (2003) Sniffers, buzzers, toggles and blinkers: dynamics of regulatory and signaling pathways in the cell. *Curr. Opin. Cell Biol.* **15**: 221–231.

Uetz, P., Giot, L., Cagney, G. *et al.* (2000) A comprehensive analysis of protein-protein interactions in *Sacchromyces cerevisiae*. *Nature* **403**: 623–627.

Wall, M.E., Dunlop, M.J. & Hlavacek, W.S. (2005) Multiple functions of a feed-forward-loop. *J. Mol. Biol.* **349**: 501–514.

Zak, D.E., Gonye, G.E., Schwaber, J.S. & Doyle III, F.J. (2003) Importance of input
   perturbations and stochastic gene expression in the reverse engineering of genetic
   regulatory networks: Insights from an identifiability analysis of an in silico network.
   Genome Res. **13:** 2396–2405.
Zhu, X., Gerstein, M. & Snyder, M. (2007) Getting connected: Analysis and principles
   of biological networks. *Genes Dev.* **21:** 1010–1024.

6

Gene regulatory network models: A dynamic and integrative approach to development

Elena R. Alvarez-Buylla, Enrique Balleza,
Mariana Benítez, Carlos Espinosa-Soto and Pablo
Padilla-Longoria

As molecular data accumulates, the use of mathematical and computational tools for understanding how the concerted actions of genes underlie developmental processes and phenotypical traits is becoming both necessary and possible. This has stimulated network theory, as a tool for understanding complex systems comprised of many connected elements with correlated behaviours and non-linear interactions. In this chapter we first present the concept of basic dynamic gene regulatory network (GRN) models and discuss which mathematical tools can be used to integrate data from complex biological processes at different space-time scales. We then explain key concepts of dynamic systems as exemplified by the Boolean case. In the following section we discuss two concepts basic to understanding GRNs of developmental processes: epistasis and robustness. After that we review work on two main approaches to studying GRN in animal and plant development. The first approach focuses on modules or subnetworks, in which behaviour is relatively autonomous and for which dynamical analyses with direct functional and/or structural interpretations are possible. By reviewing a repertoire of such small networks, we point to some generalities that are starting to emerge concerning their robust dynamic behaviour in the face of environmental and genetic perturbations. Detailed dynamical studies of specific modules enable detection of holes in experimental data and lead to novel predictions that may be tested experimentally and then fed back to refine the models. Furthermore, the application of the modular approach to understanding the genetic basis of body plan evolution in plants and animals has allowed us to hypothesize that a number of regulatory processes underlying developmental problems have evolutionarily conserved solutions. The second approach aims at recovering the complete GRN for an organism. Such studies rely on different inference methods to reverse engineer the GRN structure from genomic-wide expression arrays from different genetic backgrounds and under different environmental conditions. Dynamic analyses of genome-level GRN are still a challenge that lies ahead in both the experimental and theoretical-computational research fronts.

1 Gene regulatory network models: Are they useful for understanding development?

We still do not fully understand how genetic information determines the phenotypic traits of multicellular organisms, which is fundamental in order to comprehend development and evolution (Lewontin, 1974). The complete genome sequences of such organisms (e.g. *Arabidopsis thaliana, Caenorhabditis elegans, Mus musculus* and *Homo sapiens.* See NCBI database) are now available and demonstrate that the information stored in them is not enough to encode for the outstanding complexity of living organisms.

This vast amount of molecular-level information, as well as detailed functional analyses of single genes, now enables and requires integrative tools such as dynamical network models. These tools allow us to explore the concerted action of numerous molecular components, their interactions, and the resulting regulatory networks on the development of complex structures, functional systems or behaviours in plants and animals. Such models based on experimental data will become instrumental in testing previous theoretical predictions (Kauffman, 1969; Waddington, 1957).

These models may capture complex dynamical and structural aspects of gene regulation and aid in understanding the consequences of non-linear interactions among molecular components. In contrast with the schematic representations that are abundant in the molecular genetics literature, network models that allow simulations of dynamical aspects of a set of interacting genes may also help identify contradictions or holes in experimental data that escape intuition.

2 Mathematical tools for integrating biological processes at different time-space scales: Key for understanding pattern formation

Cell differentiation, pattern formation and morphogenesis are the main aspects of a developmental process. These are very intricate processes which involve the integration of information at different levels (*Figure 1*). Chemical reactions are taking place all the time inside and outside the cell, and the cell cycle and metabolic processes are regulated by very subtle dynamics. At the gene regulatory level (Box 1), complex networks are affected and transduce signals in the form of concentrations (or gradients of concentrations) of proteins, transcription factors and hormones, temperature variations and other environmental variables. It is clear that biochemical paths are coupled with genetic networks, influencing each other in a highly non-linear and correlated way.

At a higher hierarchical level, intercellular communication and other phenomena, such as active transport or diffusion of nutrients or other chemicals, mechanical deformations and even piezoelectric effects, play an important role in the determination of cell fate and differentiation. Development is the ultimate outcome of all these interactions.

The different signals mentioned above vary not only in their origin (chemical, mechanical, etc.), but also in terms of their specificity, duration, range of influence, speed of propagation and so on. For instance, proteins can be very specific in terms of the function they regulate and corresponding information they convey. Diffusive mechanisms are basically local and slow, whereas active transport is more global and faster.

(A)

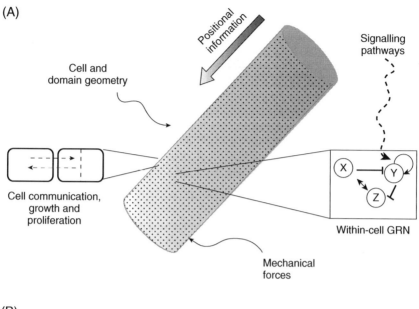

(B)

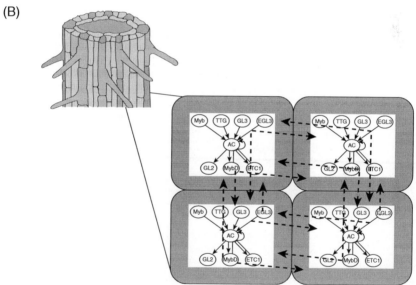

Figure 1. *Integrating information in biological systems. (A) Morphogenetic models should consider several genetic and non-genetic aspects. (B) Example of a morphogenetic model: cell fate determination in the epidermis of Arabidopsis. In this system, an intracellular GRN, mobile elements (dotted arrows), positional information (darker shaded regions in the root), geometrical characteristics of the root and other factors contribute to the formation of banded cellular patterns (see Mendoza and Alvarez-Buylla, 1998; Benítez et al., 2007 for details).*

Box 1. *From arrows to genes: Molecular processes that underlie gene regulation.*

Many of the differences among cell types and cell cycle stages can be traced to spatio-temporal
   differential gene expression. Although GRN models simplify interactions among genes as arrows,
   inputs in logic rules or variables and parameters, the molecular processes involved in gene
   regulation are very diverse and complex. Here we summarize some of these.

The events preceding gene expression and the synthesis of an active protein can be numerous and
   complex, and usually these steps and processes are precisely regulated in many ways. Gene
   regulation events are frequently characterized by the stage in gene expression at which they occur.
   Transcriptional regulation controls the synthesis of messenger RNA from a DNA template and it
   often involves specific *transcription factor* (*TF*) **complexes,** which are needed to initiate the
   polymerization of messenger RNA (mRNA) at the *promoter* region. Other genes in turn encode
   TFs and their availability greatly determines the amount of mRNA transcribed. Transcription into
   mRNA can be inhibited by proteins that bind to the DNA sequence and obstruct the polymerase
   (the enzyme that synthesizes mRNA) or prevent the recruitment of TFs.

In brief, there are *activators* that release inhibitory proteins from the DNA sequence or facilitate the
   activity of the polymerase and other associated factors, promoting gene expression; and *repressors*
   that slow or impede these events. Certain activators, the *enhancers,* promote the expression of
   genes that may be thousands of base-pairs away, probably by allowing for a particular DNA
   secondary structure that permits or speeds transcription.

DNA, in association with proteins called histones, is arranged in a highly organized structure named
   chromatin, which can exhibit two conformations: tightly packed (*heterochromatin*) or relaxed
   (*euchromatin*). Chromatin structural changes can be induced in the vicinity of a gene, facilitating its
   expression in relaxed chromatin state, because only then is it possible for the translation machinery to
   access a DNA sequence. These structural changes are mediated by numerous mechanisms, for example
   DNA methylation or histone modification, and are generically referred to as *epigenetic regulation*.
   This kind of regulation may be mediated by a great diversity of molecular factors. Interestingly,
   epigenetic regulation is responsible for gene silencing, for example of developmental genes in adult
   stages, and can be inherited by daughter cells after mitosis (Henderson and Jacobsen, 2007).

Although transcriptional regulation is the most documented and studied, there are various types of
   post-transcriptional and *translational regulation*. In eukaryotes, messenger RNA is edited before being
   translated. It is spliced (introns are removed) and the termini are modified with a 5′ cap and a 3′ poly-A
   tail. The efficiency of these processes, as well as the messenger RNA degradation rate, are clearly
   correlated with the final amount of functional protein and are also affected by the expression of other
   genes and *small RNA* sequences. Regarding the RNA sequences, it has been recently shown that a
   great number of genes are regulated, mostly in a negative way, by small RNA sequences (around
   21 nucleotides) that are complementary to some region of an mRNA. This complementarity allows for
   the small RNA to bind and promote cleavage of the mRNA, thus inhibiting gene expression.

After the protein is formed, its tagging, transport and, eventually, degradation, also influence its
   effective activity and can be precisely regulated in space and time. Finally, many proteins function
   in complexes, either with additional copies of themselves or with different proteins, and their
   activity can be different depending on which proteins they interact with. Therefore, the availability
   of different *interacting proteins* is another means of regulation.

The best mathematical representation to formalize the complex logic of regulation at different levels is the
   subject of strong debate. However, recent data suggests that at the transcriptional level, for example, a
   Boolean or discrete formalism that recovers the state of activation of a gene as one of a few possible
   discrete values (0 and 1 for the Boolean case) may be adequate (see references below). Indeed cell by
   cell analyses of transcriptional regulation dynamics suggest that this is of a digital and stochastic nature.
   The probability that a particular template is active within a certain time window, rather than the rate of
   transcription from this template, is subject to regulation. Hence, genes within individual cells have a

---

Box 1. *From arrows to genes: Molecular processes that underlie gene regulation—cont'd.*

---

distinct probability of responding to a given concentration of stimulus of transcription and the gene is either 'ON' or 'OFF' in a particular time window (Blake *et al.*, 2003; Elowitz *et al.*, 2002; Fiering *et al.*, 2000; Hume, 2000; Ozbudak *et al.*, 2002; Paulsson, 2004; Ross *et al.*, 1994). Rossi and collaborators (2000) proposed a mechanism for the observed binary response in inducible gene expression which implies competition among transcription factors with opposing functions (activation and inhibition) for the same target promoter region.

For further reading concerning molecular genetics, consult Alberts and collaborators (2002), Tijsterman and collaborators (2002), Berger (2007), Lewin (2007) and references therein.

This diversity naturally leads to the consideration of several space-time scales. In fact, it might be argued that, in contrast to many physical processes in which a relevant scale can be singled out, in biological systems all the above mentioned scales interact in a non-trivial way. From the mathematical point of view, integrating all these interactions can only be taken into account by implementing hybrid models, which incorporate both discrete and continuous elements in time as well as in space.

Another important distinction among modelled phenomena lies in the fact that, depending on the specific space-time scale with which a process is being observed, it might appear deterministic or random. For instance, Brownian motion described by a relatively big individual particle immersed in a liquid can be suitably described by means of a random walk and in quantitative terms using a Langevin equation (a stochastic differential equation), which expresses Newton's second law when random perturbations are present:

$$dv = f(x)dt + dB$$

However, the same fact, when a collection of particles at larger space-time scale is considered, can be perfectly well described by a deterministic tool, the diffusion equation:

$$\frac{\partial u}{\partial t} = D\Delta u$$

In what follows we outline some of the models in which several scales are present. Starting from experimental data, a Boolean network can be constructed in order to understand the architecture and dynamics of a genetic system (see detailed discussion below). In a more realistic scenario, one should incorporate a random perturbation. A discrete model (a Markov chain) or a continuous one (Langevin or Fokker Planck equations) can be used and the appropriate choice is again dependent on the question to be addressed. Depending on the space-time scale we are interested in, the next step could be to couple several of these models, each one valid for a cell, in a spatio-temporal context. For instance, if the system investigated is constituted by a relatively small number of cells, this can be done perfectly well by means of cellular automata.

In contrast, if we are interested in a more macroscopic perspective, for instance, the understanding of the functioning of a tissue, each individual cell loses importance, so to speak, and a continuum description can in principle be more useful. In this case, a natural mathematical tool would be a system of partial differential equations. Another advantage of the continuous approach is that other important effects can be incorporated, namely chemical reactions, active transport, diffusion, and elasto-mechanical phenomena, among others. The form these equations take in a typical example, where the diffusion and reaction of two chemicals whose concentration is given by u and v respectively are considered:

$$\frac{\partial u}{\partial t} = D\Delta u + f(u,v)$$

$$\frac{\partial v}{\partial t} = D\Delta v + g(u,v)$$

Here f and g represent the way in which the two substances interact chemically. This kind of system was first proposed by A. Turing (1952) and since then has been widely used to model morphogeneic processes, among others (e.g. Meinhardt and Gierer, 2000; Sick *et al.*, 2006). To conclude this section it is important to mention that other geometric and dynamic effects can also play an important role in pattern formation. We are referring to aspects such as growth, either due to cell growth itself or cell pro-liferation. The interplay between these two is non-trivial and related to complex issues associated in turn with other genetic control processes and developmental questions. Also the shape and size of the domain where the pattern formation process is taking place can give rise to different structures (Plaza *et al.*, 2005). For instance, it is very likely that different phylotaxes are the result of different growth rates and initial geometries (Plaza *et al.*, 2005).

3 Dynamic gene regulatory network models: The Boolean case

As discussed above, several formalisms may be used to model complex biological sys-tems, in particular GRN. As an introduction to basic terminology associated with dynamic GRN models, we will explain deterministic Boolean networks (Kauffman, 1969), which are the most intuitive (*Figure 2*). In these models, the nodes or network elements stand for genes, RNA, proteins or complexes that take part in gene regula-tion; the edges correspond to the positive (activation) or negative (inhibition) regulatory interactions among them (*Figure 2*).

In a given time t, nodes can be in one of two activation states (0, 'off' or 1, 'on'), depending on the state of other elements in the network at time $t - 1$. The way in which the state of each element is determined by other nodes is defined by updating logical rules that govern the dynamics of the system. In GRNs based on empirical information, the logical rules are grounded on experimental results (e.g. gene expres-sion patterns, loss and gain of function phenotypes, pharmacological treatments, pro-tein interaction assays, and chromatin behaviour. See also Box 1). In general, the nodes are represented as discrete variables k_1, k_2, k_3, ..., k_n and their activation state is given by rules of the general form:

$$k_i(t + 1) = F(k_{i1}(t), k_{i2}(t), ..., k_{in}(t)),$$

where F is a Boolean function and k_i can be 0 or 1.

(A)

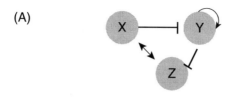

(B)

X(t)	Y(t)	Z(t)	X(t+1)	Y(t+1)	Z(t+1)
0	0	0	0	1	0
0	0	1	1	1	0
0	1	0	0	1	0
1	0	0	1	0	1
0	1	1	1	1	0
1	0	1	1	0	1
1	1	0	1	0	1
1	1	1	1	0	1

(C)

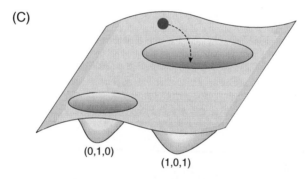

(0,1,0)

(1,0,1)

Figure 2. *Boolean gene network model. (A) Example of a three-element gene regulatory Boolean network, where X, Y and Z stand for genes, arrows for positive regulatory interactions and barred lines for negative ones. (B) Logical rules for each gene network element in (A); note that (0,1,0) and (1,0,1) are fixed point attractors of the system. (C) Schematic epigenetic landscape associated to the network: there are two attractors with basins of a size proportional to the number of gene activation state configurations or initial conditions that lead to each attractor.*

Given the set of updating rules, it is possible to follow the trajectory of any gene activation profile until it reaches a steady state or attractor, be it fixed or periodic. The number of possible configurations is 2^n where n is the number of genes in the network, and from these configurations, the ones that end up at each of the attractors constitute their basins of attraction.

It has been suggested that steady states in gene regulatory networks correspond to sustained gene activation profiles characteristic of particular cell types (Kauffman 1969; Ribeiro and Kauffman, 2007 and *Figure 3*). This is why this kind of model has

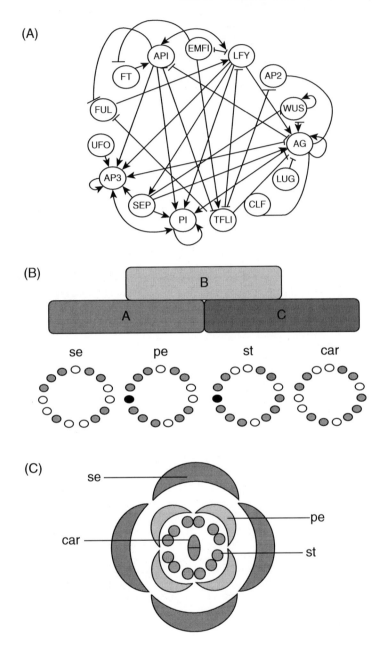

Figure 3. *Gene regulatory network model for cell fate determination in primordial organs during early flower development in* Arabidopsis *(Espinosa-Soto et al., 2004; Chaos et al., 2006). (A) Gene regulatory network grounded on experimental data. (B) Gene activity combinations that correspond to each cell type (grey genes are 'ON', white genes are 'OFF' and the state of black genes may be either 'ON' or 'OFF'): sepals, petals, stamens and carpels. These combinations are congruent with the ABC model of flower development (Coen and Meyerowitz, 1991). (C) Floral diagram showing the four types of floral organs in wild type* Arabidopsis *flowers showing the sterotypical arrangement of the great majority of flowering plant species.*

been widely explored in the context of developmental biology and, more recently, of complex diseases such as cancer (Huang and Ingber, 2006; Qi *et al.*, 2007).

A frequently used metaphor for understanding the dynamics of the multidimensional dynamics of GRNs in three dimensions is that of a landscape, in which the bottom of valleys correspond to steady states and the valley's basins to all the initial conditions that lead to each of the steady states or attractors. Every point on the landscape stands for a gene activity configuration and the shape of the valley (number, identity and relative position of attractors), is determined by the logical rules and the network's structure. In this metaphor, the system (a ball) can be located in any point of the landscape and, given the topology, it would role until a steady state is reached (*Figure 2*). Such a metaphor and the concept of epigenetic landscape were introduced by C. H. Waddington (1957).

Because of the lack of experimental information, it is often assumed that all the elements in a network are updated synchronically, that is, their states are updated at the same time. However, this might not be the case in real genetic systems. Moreover, the updating regime can induce changes in the identity and number of attractors when these are periodic, and on the size of attractive basins (Gershenson, 2002). This is why, when modelling a GRN, different updating regimes should be tested. Also, it should be kept in mind that, even though these models may offer great insight concerning the identity of steady states and other qualitative aspects of the system, they might not be useful for predicting precise trajectories of a system, unless there is enough empirical information to feed into the model.

It is worth mentioning that most biological systems that have been modelled with GRN exhibit a greater number of fixed point attractors than expected for GRN of similar size and connectivity in networks with random logical rules, suggesting that biological networks are biased towards a relatively ordered and rich (in terms of the number of attractors for a GRN of a certain size) behaviour and that such bias relies on the nature of the updating rules. This is an issue that, although already explored by some authors (e.g. Shmulevich *et al.*, 2003), remains unclear.

For a long time Boolean GRN were considered too simplistic to describe regulatory processes. However, recent studies in both plant and animal systems (reviewed below) have demonstrated that this relatively simple formalism may be quite useful for modelling the logic of gene regulation during cell differentiation. Recent theoretical studies (Klemm and Bornholdt, 2005a, 2005b) have shown that Boolean networks recover all robust attractors of equivalent continuous systems. Such attractors may be the ones that are biologically relevant and thus may explain the success of Boolean GRN in recovering the steady-state gene activation arrays observed in biological systems studied. Furthermore, the logic of regulation, at least at the transcriptional level, seems to be suitably represented with a Boolean function (see Box 1).

4 Epistasis and robustness: Two sides of the same coin for understanding developmental constraints?

Genes interact in complex regulatory networks, and thus the effect of several genes in the construction of the phenotype is not additive; the contribution of a gene to a phenotypic trait depends on its genetic background. This kind of relationship of dependence and non-additivity in gene effects is known as epistasis.

Epistasis is the usual condition in gene activity since even 'simple Mendelian traits', such as the sickle-cell disease, have been shown to be affected by genetic background (Templeton, 2000). Epistasis is also responsible for the non-linearity in the relationship between genotype and phenotype (Rice, 2000). Hence, to address the problem of phenotypic evolution, it is necessary to consider epistasis.

The set of epistatic relationships among genes, along with other non-genetic factors, comprise development and limit the number of possible phenotypical variants. This imposes a constraint to the action of natural selection: some of the phenotypes we do not find in nature are missing because, given existing developmental mechanisms, it is unlikely (or impossible) to produce them (see Beldade and Brakefield, 2003 and Arthur, 2003 as examples of the controversy about the importance of these constraints and natural selection). These restrictions are referred to as developmental constraints, and were defined as: 'a bias on the production of variant phenotypes or a limitation on phenotypic variability caused by the structure, character, composition, or dynamics of the developmental system' (Maynard Smith et al., 1985).

Phenotypic robustness, phenotype constancy in the face of heritable or transient (environmental) perturbations, is an important kind of developmental constraint. As with all other developmental constraints, phenotypic robustness is a result of gene interaction. In fact, epistasis is a necessary condition for phenotypic robustness (Rice, 2000) that is observed in very different kinds of traits: gene expression, RNA, protein structure and folding, morphological traits, or fitness (de Visser et al., 2003).

Phenotypic robustness may be observed in the face of either heritable or non-heritable perturbations. Genetic robustness is that observed in the face of heritable alterations, such as mutations. On the other hand, environmental robustness is the phenotypic robustness in the face of non-heritable, therefore transient, perturbations (de Visser et al., 2003; Rice, 2000). Non-heritable transient perturbations may be the result of external (humidity, temperature, etc.) or internal factors, such as developmental noise that can arise as a consequence of thermal fluctuations or 'sampling errors' because the number of molecules of each kind is finite within cells. Developmental noise may produce stochastic changes in a developmental trajectory. In a model of a particular developmental system, transient perturbations can be represented as changes in the system state, and hence in the value of some of the variables (e.g. protein concentration or gene expression state), while heritable perturbations may be modelled as changes in the system definition, for instance in exponents, parameter values, and so on. In the Boolean GRN model introduced above, an example of a transient alteration would be a change in the initial conditions of the system, while a genetic alteration could be simulated by permanently turning a gene 'off' or permanently altering the logical rule of one or more nodes.

If a developmental mechanism has a high environmental robustness, then it can perform certain functions relatively independently of other processes within the organism (von Dassow and Munro, 1999), and hence function as a semi-autonomous module. Modularity promotes evolvability as it permits the combination of semi-autonomous functions without significantly altering the basic original behaviour of each module. Phenotypic robustness is also responsible for the repeatability of developmental traits among individuals of the same species. Only if the generation of a phenotypic trait is robust can this trait appear in different genetically identical organisms under various environmental conditions. On the other hand, robustness to genetic alterations yields repeatable developmental patterns under diverse genetic backgrounds, or even among different species. Robustness enables continuity in the presence of a phenotypic trait along a lineage and, therefore, is a requisite for the preservation of homologous traits. The more robust a developmental module is, the more we expect conservation of traits regulated by this module along wider or more distantly related taxa.

Some of the modules underlying relatively conserved aspects of body plan patterning in plants and animals have been studied in both using dynamical models of GRN.

5 Dynamic GRN models for animal and plant modules

5.1 Cell patterning

We still lack the computation power or mathematical formalisms that can keep track of the dynamics of the global GRN for even the simplest organism. An example is *E. coli*, for which the complete GRN has in the order of 4000 genes (http://regulondb. ccg.unam.mx/). However, an underlying modularity of the organism's global GRN enables studying development as a collection of processes that can be logically isolated and modelled (Schlosser and Wagner, 2004; Valente and Cusick, 2006). This enables focusing on gene sub-networks, which can be studied as dynamic systems, corresponding to modules that are structurally and functionally isolated from the rest and that have been thoroughly studied in terms of molecular genetics.

Odell's laboratory (von Dassow and Odell, 2002: von Dassow *et al.*, 2000) proposed one of the first GRN dynamic models for such a developmental model. Its analysis showed that the *Drosophila* gene network that regulates patterning during early fly development and sets the polarity patterns in the embryo's segments is robust for different initial conditions or parameter values. These affect, among others, the strength of interactions and the exact form of the 'genes or proteins' activation functions (von Dassow *et al.*, 2000). Such transient robustness supports the idea that the studied GRN performs its function semi-autonomously irrespective of its interactions with other genes outside the module studied (von Dassow and Munro, 1999). Furthermore, a Boolean model of the same GRN (Albert and Othmer, 2003) recovers the same patterns for the *Drosophila* segment polarity genes as those recovered by the continuous model (von Dassow *et al.*, 2000).

Another module studied by the same group using dynamic GRN models is one that regulates the neurogenic and proneural patterns in *Drosophila* (Meir *et al.*, 2002). Interestingly, in this case the module studied also shows a robust behaviour. Furthermore, in this case structural alterations, and not only parameter changes, are tolerated, thus constituting a case of genetic robustness.

In plants, a few developmental modules have been studied using dynamic GRN models. The first Boolean GRN model grounded on experimental data that was shown to recover observed gene activation profiles was proposed for the module that underlies the combinatorial gene activation profiles that characterize each of the four floral primordia regions early in flower development. These eventually give rise to each of the floral organs that characterize the widely fixed floral plan of most higher flowering plant species (eudicotyledonous angiosperms; Rudall, 1987): sepals, petals, stamens and carpels.

This floral GRN model was grounded on data for the experimental plant *Arabidopsis thaliana* (Mendoza and Alvarez-Buylla, 1998, 2000; *Figure 3*) and constituted a Boolean model for single cells based on available experimental data. Interestingly, the same GRN was analysed using the Thomas (1991) logical approach, that is based on identifying the functional positive and negative feedback loops. The results coincided with predictions made by the now classical ABC model of floral organ specification (Coen and Meyerowitz, 1991), as well as the preceding analyses of the same GRN (Mendoza and Alvarez-Buylla, 1998).

The Thomas (1991) approach has also been applied to analyse a GRN important for *Drosophila melanogaster* development (Sánchez and Thieffry, 2001). Actually, this approach can also be used to identify functional modules. In the case of the floral GRN, for example, two main modules are identified: one that underlies the 'A or C' decision and another one that underlies the B function (Thieffry and Sánchez, 2004).

More recent experimental data has been used to update the GRN floral model and this still shows that all initial conditions converge to gene activity states that match expression profiles of primordial cells of sepals, petals, stamens and carpels, and genetic perturbations of this GNR reproduce known mutant patterns (Chaos *et al.*, 2006; Espinosa-Soto *et al.*, 2004; *Figure 3*). These results suggest that this GRN incorporates the key components of the module responsible for the combinatorial gene activity that underlies the ABC model and provides a dynamical explanation for this model.

As in the *Drosophila* study from Odell's laboratory, our analyses showed that the steady states of the floral GRN are robust to initial conditions, but we also found that they are robust to changes in the interaction rules (Chaos *et al.*, 2006; Espinosa-Soto *et al.*, 2004), as was the case for the neurogenic module (Meir *et al.*, 2002). Dynamic models for well-documented sub-networks are also useful to detect inconsistencies or holes in experimental data and in order to propose novel predictions. Indeed, we predicted, for example (Espinosa-Soto *et al.*, 2004; see Mendoza and Alvarez-Buylla, 1998; and Mendoza & Alvarez-Buylla, 1999 for other examples), that the gene *AGAMOUS* should self-activate and this was confirmed in an independent parallel experimental paper of R. Sablowsky's laboratory (Gómez-Mena *et al.*, 2005).

Another module that has been analysed for *A. thaliana* is the one that underlies cell-type specification in the epidermis of aerial and root tissues (reviewed by Pesch and Hülskamp, 2004; *Figure 1B*). The epidermis of both leaves and roots exhibit two basic types of cells: with and without hairs. However, the spatial patterns of these two cell types are contrasting, even though the genetic networks regulating their respective cell-fate determination have very similar structures and components (Benítez *et al.*, 2007; Mendoza and Alvarez-Buylla, 1998). We have recently proposed a dynamic GRN model in order to integrate available experimental data for leaf and root hair patterning and have shown that such GRN models recover the two observed cell types. Furthermore, we argued that the GRN could be reduced to activator-inhibitor systems that recover previously expected results for these kinds of dynamic models, including striped and dotted cell patterns found in the root and leaf epidermis of *A. thaliana*, respectively.

The work on *A. thaliana* epidermal cell patterning illustrates that simple formal tools can lead to novel insights on current data and enable precise hypotheses that can be addressed experimentally. Despite the fact that the coupled spatial-temporal dynamics of the complete GRN underlying such cell patterning has yet to be studied, the simple models that have been put forward already show that despite subtle differences in the root and leaf networks, they have equivalent dynamical behaviours. Our results also suggest what type of biasing signal may underlie the characteristic striped cell patterning of hair cells in the root of *A. thaliana* in contrast to the almost uniform distribution of hair cells in the leaf epidermis.

Our simulations also indicate that other types of developmental constraints, such as cell shape, may affect pattern stability in the root. This study thus supports the idea that in this and other cases, contrasting spatial cell patterns and other evolutionary morphogenetic novelties may originate from conserved genetic network modules subject to divergent contextual traits.

The studies reviewed here and other similar ones, suggest that in order to understand how the complete and huge biological GRN have been assembled during biological evolution, the identification and dynamical analysis of structural-functional (and historical) modules will be necessary. For example, Arenas and collaborators (2006) have implemented a method to analyse a GRN's dynamics by studying how different nodes are coupled. Similar approaches using stochastic systems have also been proposed to study which GRN nodes may be critical in complex diseases such as cancer (Shmulevich *et al.*, 2002).

Recent papers have put forward another level of GRN structuring at an even smaller level. This implies identifying small motifs of a relatively small number of nodes within modules or large GRN. Analyses are done to see if there are particular unions among different numbers of network nodes that are over-represented with respect to what is expected for randomly assembled GRN. The existence of such 'motifs' can be useful for classifying the modules or larger GRN into different types and provides clues with respect to the type of information processing that each module or collection of motifs is able to do. It is thus a useful approach to start linking GRN and module structure to function.

The identification of motifs has been put forward by Milo and collaborators (2004) who analysed Internet, regulatory, language and food web networks. They found that these networks are not assembled randomly and specific motifs characterize each network/module or are under-represented in them.

Since GRN are perhaps more likely to be proposed by biologists, who are expected to be the best trained to formalize into logical rules the available experimental data, more practical and 'user friendly' tools are required. Examples of such software, for the construction of Boolean GRN, have recently been released (DDLab, Wuensche, 2001; GINsim, González *et al.*, 2006), but these have to be tested and improved. Additionally, a community-based strategy to integrate available experimental data and formalize it in the form of logical functions or in other ways should be implemented.

It is probably too soon to find generalities among biological GRN underlying functional modules. However, it seems, for example, that in the topology of biological GRN grounded on experimental data, the degree of distribution of incoming connections is of Poisson type and that of the outgoing connections is of a 'scale free' type. This suggests that most genes are regulated by few genes and only a few are global regulators (reviewed by Albert, 2005).

The use of GRN dynamic models to integrate molecular genetic data on plant and animal development is also enabling researchers to identify generic aspects of development. It seems that, in contrast to what was previously believed, precise signalling pathways are not required to determine cell types, as these are established as a dynamic consequence of overall GRN topology. Interestingly, as more GRN and regulatory motifs are documented experimentally, the prediction of former theoreticians that a limited number of different mechanisms capable of generating and maintaining heterogeneities during morphogenesis would be found among living organisms, is being fulfilled. The so called activator-inhibitor system (Meinhardt and Gierer, 2000) that is repeatedly found to underlie various patterning processes of phylogenetically distant taxa is an example.

It is also becoming apparent that particular topological characteristics of GRN, such as the presence of feedback loops, could underlie the generic robustness of documented biological GRN. This had already been suggested by theoretical studies (e.g. Hogeweg, 2000). However, we are still far from being able to postulate general trends in the interplay between structure and function of biological GRN underlying different types of developmental processes. The postulation of such general behaviours will most likely stem from analyses of well-documented modules which have clear functional and evolutionary interpretations. Nonetheless, both the study of relatively small modules as well as efforts to infer GRN at larger genomic scales should feedback to each other. More general theoretical frameworks should be kept in mind as well (Alvarez-Buylla *et al.*, 2007). Such combined analyses will eventually lead to uncovering general principles for network assemblage, dynamics and pattern formation.

5.2 *Body plan development and evolution*

A central aim of modelling and theorizing in biology is to state and, at least partially, answer relevant evolutionary issues. Gene regulatory network models, along with other integrative dynamic models, allow us to address both generic and particular questions about the origin of patterns during development and, importantly, the evolution of such patterns, forms and organisms within lineages (see, for example, Damen, 2007; Davidson and Erwin, 2006; Wagner, 2007).

These models have already contributed to the discovery of potential generic properties regarding the structure and dynamics of diverse biological networks, such as those constituted by genes, populations, signalling pathways, and so on (Almaas, 2007). They have also pointed to promising avenues in the understanding of complex diseases (Huang and Ingber, 2006; Qi *et al.*, 2007) and, in general, have encouraged the statement of questions, experiments and explanations that might not have arisen in another context.

However, so far, few morphogenetic models have incorporated experimentally grounded GRN. Most models addressing the origin of cellular and organismal patterns consider 'toy networks', while the majority of experimentally grounded GRN models ignore cellular-scale interactions. Nevertheless, the emergence and maintenance of cellular arrangements is affected in complex ways by several genetic and non-genetic elements. For instance, positional information (Dolan, 2006), geometry of the domain (Diambra and da Fontoura Costa, 2006), mechanical restrictions (Dumais, 2007) and hormonal and environmental factors (e.g. Jackson *et al.*, 2002; Jonsson *et al.*, 2006; Wang *et al.*, 2007) all play relevant roles in this process.

Therefore, one of the challenges that remains consists of developing multi-scale models, most likely by the postulation of hybrid models of continuous, discrete, stochastic and deterministic types, that integrate gene regulatory networks in a cellular context with the elements mentioned above, among others, and permit a better understanding of organismal level morphogenesis and evolution.

Comparative approaches that in their broadest sense should include both plants and animals are important for searching for generalities (Meyerowitz, 2002). Correlations of structural and dynamical aspects of GRN with variants of the morphological traits regulated by such networks are being explored (Abouheif and Wray, 2002; Davidson and Erwin, 2006; Espinosa-Soto et al., 2004). For example, von Dassow and collaborators (von Dassow and Munro 1999; von Dassow et al., 2000) suggested that the robustness and alterations of the segment polarity gene network in insects could underlie the overall conservation of body plan and the origin of long and short germ-band insects, respectively. Similarly, the GRN model that we have proposed for primordial cell type specification during flower development (Espinosa-Soto et al., 2004) behaves as a robust module and provides a possible explanation for the conserved basic floral plan of eudicot plants. However, these theoretical frameworks could also be used to understand observed divergent phenotypes such as the unique floral arrangement of the inside-out flower of the endemic Mexican plant, *Lacandonia schismatica* (Ambrose et al., 2006; Vergara-Silva et al., 2003).

6 Genome level GRN models from microarray data: A challenge still ahead

Genomic approaches aim at inferring the organism's complete GRN. To this end both automated experimental and statistical methods are actively developed and tested (see, for example, reviews in: Affymetrix, 2004; Ball et al., 2005; Quackenbush, 2001). The basis for inferring or reverse engineering GRN structure consists of two general approaches: correlate gene changes in gene expression data or find GRN models (with different imposed restrictions) that explain the observed data. For the inference to work for a particular organism, it is essential to have the expression data for the full complement of genes, or as near as, under a variety of different environmental or genetic conditions. There are several public databases where it is possible to find hundreds of microarray experiments from different organisms.[1] In order to clarify how data is processed and GRN structure inferred from it, we use a small five-gene data subset from the *Bacillus subtilis* whole genome microarray retrieved from www.genome.jp/kegg/expression and exemplify the two most widely used inference methods: 'Bayesian' and 'Mutual Information' (see Appendix).

Networks obtained by these two methods have been extensively validated with experimental data (examples of Bayesian inference in Friedman, 2004; Segal et al., 2003). In this method, different GRN structures are proposed and are scored based on how well they explain the observed microarray data. A Bayesian GRN is an acyclical graphical representation of a joint probability distribution where the nodes are the random variables and the directed edges are causal influences. The second method that we exemplify uses Mutual Information (MI) to decide if two gene expression patterns from two different genes are independent. An interaction (with no direction) is

[1] For example, visit the Stanford Microarray Database http://genome-www5.stanford.edu

inferred if the MI of the expression patterns is significantly higher than the MI of a random shuffling of the same patterns (Liang *et al.*, 1998; Steuer *et al.*, 2002; example in Basso *et al.*, 2005).

Other GRN structure inference methods have been proposed. For example the one first used for an *Arabidopsis thaliana* GRN model (Wang *et al.*, 2006), assumed that GRN operate near a steady state. The assumption justifies approximating the dynamics of GRN with systems of linear differential equations with undetermined rate constants. The system is reverse engineered (the rate constants in the linear model determined) by fitting the rate constants to the microarray data using a linear algebra technique known as Singular Value Decomposition. Other methods that require more detailed information can be found in Aracena and Demongeot (2004) and Perkins and collaborators (2006). These methods may be useful links between more carefully studied networks and other reverse GRN structure inference methods in a recursive process in which functional data is considered to propose and validate regulatory interactions inferred from microarray data.

While the functional genomic approaches aim at a full account of an organism's GRN, the experimental and simulation tools that are needed to have a complete dynamical account of a whole organism's GRN are still far away. However, insights from structure-function analyses in a large enough repertoire of experimentally grounded GRN for well characterized modules will contribute in this direction.

Appendix: Gene network inference methods: From raw microarray data to a GRN model

The data set we selected to demonstrate inference methods is particularly suitable because all microarrays were synthesized by only one laboratory, 96% of the genes in *B. subtilis* were successfully amplified, microarrays were normalized (see below) and the set covers a relatively large group of deletions/overexpressions[2] experiments.

Microarray technology has been implemented in various ways (known as platforms) but the most widely used are two-colour spotted arrays and the commercial Affymetrix GeneChip. Since our *B. subtilis* set is two-colour spotted we will concentrate on this particular platform. Two-colour spotted microarrays measure the relative value of the gene expression intensities between two samples (two channels), one of them usually a control. Obtaining the data (expression intensities) from two-colour spotted microarrays used in the case under analysis implies several steps. An estimation of the noise of every spot in every channel is obtained by measuring the intensity of the signal in an area just next to the spot. This 'background' signal is subtracted from the raw intensity. Then, the total intensity in every channel is normalized by the median[3] obtained from all the intensities in the channel that is being normalized (Ball *et al.*, 2005; Quackenbush, 2001), see *Figure A1*. The assumption for this normalization to work is that a cell type, in every cellular context or in every one of its different phenotypes, expresses on average the same amount of mRNA. From the background corrected and normalized channels, log ratios (base 2 in order to directly

---

[2] 69 microarrays in total at the time of writing.
[3] Sometimes also the average is used. In fact, data from *B. subtilis* is normalized using the average.

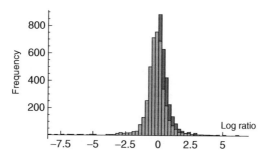

Figure A1. *Histogram of log ratios of a two-colour spotted whole genome microarray experiment (taken from Blattner Lab, http://www.genome.wisc.edu/). The experiment compares the expression changes of* Escherichia coli *between two growth conditions: aerobic and anaerobic growth. The histogram in the foreground shows log ratios of un-normalized data. The histogram in the background shows log ratios of median-normalized data. Note that the un-normalized histogram is shifted to the left reflecting an unreal whole genome under-expression.*

read fold-changes) are calculated. Data downloaded from different databases are processed at different levels.

We validate the results from the inference methods explained here using a 5 gene sub-network from the curated *B. subtilis* network (Release 4) available at http://dbtbs.hgc.jp, see *Figure A2*. Also, in *Table A1* we show the log ratios of the 5 selected genes in some of the 69 microarrays. We corrected for background noise and calculated the log ratios from the retrieved microarray data.

Bayesian Structure Learning

Bayesian Structure Learning solves the following problem: given a set of data, find the network structure that *explains the data* with the highest probability (Heckerman, 1996;

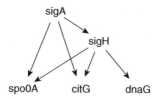

Figure A2. *A five gene sub-network taken from the curated* B. subtilis *network (Release 4). The network is arbitrary except for the fact that all the regulators (sigA, sigH) of the regulated genes spo0A, citG and dnaG are included. Note that this sub-network is embedded (contextualized) in the whole gene network of* B. subtilis *and the state of sigA and sigH depends on other genes. However, taking advantage of our knowledge of the curated network, we know that the importance of the regulators of sigA and sigH is secondary while only sigA and sigH are the direct cause of change of spo0A, citG and dnaG.*

Table A1. *An excerpt of the background corrected and normalized log ratios of* B. subtilis *microarray data.*

	ex0000263	ex0000264	ex0000265	Ex0000266	ex0000267	Ex0000268	ex0000269
SigA	−0.3741	−0.2766	−0.1834	−0.2161	0.0961	0.4485	0.1722
SigH	−0.6904	0.2034	0.4673	−0.7327	−0.1023	0.3185	−0.1978
spo0A	0.7404	0.0487	−0.2547	0.5611	−0.2154	0.6724	0.0687
CitG	−0.0245	−0.4530	0.0541	−0.5242	0.1394	0.1114	−0.2397
DnaG	−0.4996	1.8188	0	−0.5639	−0.4039	−2.0704	−1.8220

Data from seven microarray experiments, out of a total of 69, are shown. Column labels: IDs of microarrays retrieved from www.genome.jp/kegg/expression. Row labels: The gene name of the five genes composing the network in *Figure A2*. Note that the log ratio in ex0000265 of gene dnaG is zero. Every time a subtraction of the background intensity to the raw intensity gives a negative intensity we set the log ratio equal to zero (no change detected). This is a rare event and will not affect the inference because both techniques here explained are robust to noise.

Neapolitan, 2004; Pearl, 2000). To discriminate the possible structures we use a scoring function:

$$p(D|S) = \prod_{i=1}^{n} \prod_{j=1}^{q_i} \frac{\Gamma(N_{ij})}{\Gamma(N_{ij} + M_{ij})} \prod_{k=1}^{r_i} \frac{\Gamma(a_{ijk} + s_{ijk})}{\Gamma(a_{ijk})}.$$

This scoring function assumes discretized data. In the literature it is known as the Bayesian Dirichlet (BDe) scoring metric. On the left side, we have *the probability of data D given structure S*. On the right side, the first product goes over all n genes in the network. The second product goes over all different configurations, q_i, regulators can take when regulating gene i. In the third product, r_i is the number of states gene i can assume; for example if $r_i = 2$, gene i can be expressed or inhibited. s_{ijk} is the number of cases in the data such that gene i happens to be in state k when the configuration of the regulators is j. M_{ij} is the *total* number of cases in the data such that the regulators of gene i are in the j-th configuration, that is $M_{ij} = \sum_k s_{ijk}$. Think of a_{ijk} as initialization parameters, they are *a priori* cases whose values are set by you to compete in importance with the data cases a_{ijk}: if you trust your data you will set $a_{ijk} < s_{ijk}$. N_{ij} plays the same role as M_{ij}, i.e. $N_{ij} = \sum_k a_{ijk}$. We herein clarify the different terms with an example.

Imagine that we have two genes, A and B, and you are not certain if there is a regulation between them. Possible data D (assuming that genes can be in two states, 0 inhibited, 1 expressed) is:

	C_1	C_2	C_3	C_4	C_5	C_6
A	0	1	1	0	1	1
B	1	0	0	1	1	0

C_i denotes a certain cellular context, phenotypic variation, etc. (equivalent to the IDs that tag different experiments in *Table A1*). We first evaluate the structural hypothesis '*B regulates A*' ($B \rightarrow A$). This sets $q_A = 2$ because B, the only regulator of A, can take two configurations[4] 0 or 1. Given that B is a *root* (no gene regulates it) we set $q_B = 1$. Thus, counting the data we have for A: $s_{A,B=0,A=0} = 0$, $s_{A,B=0,A=1} = 3$, $s_{A,B=1,A=0} = 2$, $s_{A,B=1,A=1} = 1$. And for B: $s_{B,B=0} = 3$, $s_{B,B=1} = 3$. Note that we have suppressed the j index for B because $q_B = 1$. It is customary to take $a_{ijk} = N/r_i q_i$ with N a positive number, most of the time equal to one. With the previous setting of a_{ijk} we say we are using an *Equivalent Sample Size* (ESS), this is standard when learning structure.[5] Using $N = 1$ the score is:

$$p(D|S_1) = \left(\frac{\Gamma(\frac{1}{2}+\frac{1}{2})}{\Gamma(\frac{1}{2}+\frac{1}{2}+3+3)} \frac{\Gamma(\frac{1}{2}+3)}{\Gamma(\frac{1}{2})} \frac{\Gamma(\frac{1}{2}+3)}{\Gamma(\frac{1}{2})} \right) \left(\frac{\Gamma(\frac{1}{4}+\frac{1}{4})}{\Gamma(\frac{1}{4}+\frac{1}{4}+0+3)} \frac{\Gamma(\frac{1}{4}+0)}{\Gamma(\frac{1}{4})} \frac{\Gamma(\frac{1}{4}+3)}{\Gamma(\frac{1}{4})} \right)$$

$$\left(\frac{\Gamma(\frac{1}{4}+\frac{1}{4})}{\Gamma(\frac{1}{4}+\frac{1}{4}+2+1)} \frac{\Gamma(\frac{1}{4}+2)}{\Gamma(\frac{1}{4})} \frac{\Gamma(\frac{1}{4}+1)}{\Gamma(\frac{1}{4})} \right)$$

S_1 denotes the structure hypothesis $B \rightarrow A$. The first parenthesis in the product is the contribution of B to the score. The second parenthesis is the contribution of A when $B = 0$. The third is the contribution of A when $B = 1$.

The second structural hypothesis, no regulation between A and B, sets $q_A = 1$ and $q_B = 1$ because both genes have no regulators. Also: $s_{A,A=0} = 2$, $s_{A,A=1} = 4$ and $s_{B,B=0} = 3$, $s_{B,B=1} = 3$. Using again $N = 1$ the score is:

---

[4] If A had two regulators B and C, q_A would be equal to 4 because there would be four configurations of the regulators that could in principle affect A in different ways. The four configurations are: 00, 10, 01, 11. For example, the second one, 10, says B expressed C inhibited.

[5] Using an ESS means we initialize the network distributing the same quantity N of *a priori evidence* (*a priori* data) to all genes. We can see this calculating the *total* quantity of *a priori* evidence $\sum_j N_{ij}$ in gene i with the ESS assumption $a_{ijk} = N/r_i q_i$. First $N_{ij} = \sum_k a_{ijk} = \sum_k N/r_i q_i = N/r_i q_i(\sum_k) = N/r_i q_i(r_i) = N/q_i$. Note that the sum $\sum_k$ is equal to r_i because k is the index of the different r_i states of gene i. Now $\sum_j N_{ij} = \sum_j \sum_k a_{ijk} = \sum_j N/q_i = N/q_i(\sum_j) = N/q_i (q_i) = N$. Note again that the sum $\sum_j$ is equal to q_i because for gene i there are q_i different configurations of its regulators. Thus the total quantity of *a priori* evidence for gene i is independent of i and equal to N, i.e., we distribute the same amount of evidence, N, to all genes.

$$p(D|S_2) = \left(\frac{\Gamma(\frac{1}{2}+\frac{1}{2})}{\Gamma(\frac{1}{2}+\frac{1}{2}+3+3)} \frac{\Gamma(\frac{1}{2}+3)}{\Gamma(\frac{1}{2})} \frac{\Gamma(\frac{1}{2}+3)}{\Gamma(\frac{1}{2})} \right) \left(\frac{\Gamma(\frac{1}{2}+\frac{1}{2})}{\Gamma(\frac{1}{2}+\frac{1}{2}+2+4)} \frac{\Gamma(\frac{1}{2}+2)}{\Gamma(\frac{1}{2})} \frac{\Gamma(\frac{1}{2}+4)}{\Gamma(\frac{1}{2})} \right)$$

A numerical evaluation of the scores gives $p(D|S_1) = 7.6294 \times 10^{-5}$ and $p(D|S_2) = 3.3379 \times 10^{-5}$. As we can see, the data supports regulation between B and A. However, we must mention that if you try to calculate the score of the structure $A{\to}B$ you will obtain the same score as the one of $B \to A$. This is because the structures $A \to B$ and $B \to A$ have the same *graph pattern* (represented by A–B the edge without direction) and all members of a graph pattern have the same score, see Neapolitan (2004) to know more on graph patterns. With this *degeneracy* in mind (two networks with the same score), you may think that Bayesian Theory is not so powerful but note that, after applying the structure learning, you have the valuable information that A and B do have a relation.

Do not think we omit analysing the network structure with the loop $A{\to}B{\to}A$. Loops are not considered by Bayesian Networks; they are the realm of Dynamic Bayesian Networks. Thus, when you automate your search for the best network with a computer program it is important to know if a proposed random network has any loops in order to reject it. A very efficient way to do this is with the algorithm *Depth First Search*, DFS.[6] In short, this algorithm follows paths as far as possible in a network. When, in a search, a previously visited vertex or a vertex with no outgoing edges is reached, the algorithm backtracks to the previous vertex and restarts the search selecting another path if possible. If not, it backtracks again until there are no unvisited vertexes. Note that if there is a cycle in the network, you can finish the search without waiting for all vertexes to be visited: If you reach a previously visited vertex then you have *at least a cycle* and your search finishes; if you never reach a previously visited vertex from any path then your network is acyclic.

In the previous example we chose the network with highest score. This strategy is generally known as a *greedy search*. Thus a complete implementation of a search strategy in graph space to find the best structure given a set of data is the following: propose a random network, score it, propose some changes to the previous structure,[7] check for cycles, if no cycles are detected score the new network, if the new score is higher than the previous one save the new network, repeat until the highest score is found.

Before we infer the best network structure from our microarray data, see *Table A1*, we must first *discretize* it. Yu, using *synthetic data*, concludes that *three* categories are optimal (Yu *et al.*, 2002, 2004). You can also corroborate this by using the *real gene expression data* that we are testing here and trying out different numbers of categories. The first category (out of three) is for genes that show an under-expression, that is,

---

[6] For an implementation of the algorithm in pseudocode see Jungnickel (2002).

[7] Choosing two genes there are three possibilities: if there is a regulation between them invert it or eliminate it and if there is no regulation between them create one.

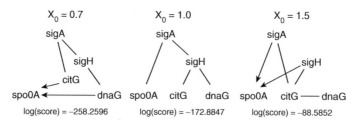

Figure A3. The best three scoring networks using three different discretization thresholds, x_0. Note that we show, rather than networks, graph patterns. To generate a particular network from any of the patterns, you have to assign a direction to the undirected edges being careful of not adding a head to head meeting ($X \rightarrow Z \rightarrow Y$) or a cycle. For example in the graph pattern with $x_0 = 0.7$ it is permissible a direction assignment cit $G \rightarrow sigA \rightarrow sigH$ contrary to $citG \rightarrow sigA \leftarrow sigH$ which creates a head-to-head meeting. Remember that all networks from a particular graph pattern will score the same. We present the log of the scores; this is quite customary in the literature.

genes with a log ratio lower than $-x_0$ (x_0 is an arbitrary real number playing the role of a threshold). In the third category we find over-expressed genes (log ratios higher than x_0). The second category is for genes that do not show a considerable change (log ratios between $-x_0$ and x_0). Note that, since we now have three categories, r_i in the scoring function is three rather than two as in the previous example. With this in mind we must discover the best value for x_0. In *Figure A3* we show the best three scoring networks (obtained doing a greedy search in graph space) using different values of x_0 to discretize the data whose excerpt we show in *Table A1*. To evaluate the integrity of our inference we count the number of different edges (regardless of direction) between the inferred networks and the real one, *Figure A2*; this is the *structural difference*. For the network with $x_0 = 0.7$ we have a structural difference of four because the inferred network has one extra edge (dnaG–spo0A) and three missing ones (sigA–spo0A, sigH–spo0A, sigH–citG). For the network with $x_0 = 1.0$ the structural difference is two. For the network with $x_0 = 1.5$ the structural difference is four. We conclude that the best data discretization threshold is $x_0 = 1.0$. Note that although we are inferring a previously known network, this is a vital step towards learning unknown networks because you have the opportunity to calibrate the different parameters in your program and characterize it with well-known sub-networks before trying to infer new regulations.

Inferring network structure using mutual information

The Mutual Information (MI) between two variables A and B is:

$$I(A,B) = \sum_i^{N_A} \sum_j^{N_B} p(a_i,b_j) \log \frac{p(a_i,b_j)}{p(a_i)p(b_j)}.$$

We can think of A and B as two genes, N_A the number of states gene A can assume and $p(a_i)$ the probability of gene A to be in state a_i. Analogously N_B, $p(b_j)$. $p(a_i,b_j)$ is the joint probability distribution. Note that $I(A,B) = 0$ when A and B are statistically independent, $p(a_i,b_j) = p(a_i)p(b_j)$ (Steuer et al., 2002).

We illustrate the calculation of the MI with the same data used in the two gene example of the previous subsection. The data is binary thus the number of states, of A and B is two. We estimate the different probabilities using the frequencies extracted from the data: $p(a_0) = 2/6, p(a_1) = 4/6, p(b_0) = 3/6, p(b_1) = 3/6, p(a_0, b_0) = 0, p(a_1, b_0) = 3/6, p(a_0, b_1) = 2/6, p(a_1, b_1) = 1/6$. With these values we obtain $I(A,B) = 0.31826$. In order to know if there is some kind of regulation between A and B we compare the value of $I(A,B)$ with the MI of a *null hypothesis*: the independence of A and B. We begin by creating an *ensemble* of surrogate data sets to represent our null hypothesis. Surrogates are obtained by randomly permuting the original data to get rid of all dependence (if any) between A and B. We reject the null hypothesis with respect to a certain *significance level S* given by:

$$S = \frac{I_{data} - \langle I_{surr} \rangle}{\sigma_{surr}}.$$

We assume that the values of the MI of the *ensemble* of surrogates are Gaussian distributed with average $\langle I_{surr} \rangle$ and standard deviation $\langle \sigma_{surr} \rangle$ We can see S is the number of standard deviations I_{data} is from $\langle I_{surr} \rangle$. For example, generating 1000 surrogates from the data of our binary example, we obtain $\langle I_{surr} \rangle = 0.12562$ and $\sigma = 0.15557$. This gives $S = 1.23831$ and represents that A and B are not independent at a significant level of 79%.[8] An edge (with no direction) between A and B is inferred if an arbitrary imposed significance threshold S_0 is surpassed.

Therefore, the general approach to infer regulatory interactions between all possible gene pairs in a GRN using MI is: measure the MI, generate an *ensemble* of surrogate data sets, calculate $\langle I_{surr} \rangle$ and σ_{surr}, use a certain significance threshold S_0 to decide if a regulatory interaction is inferred or not.

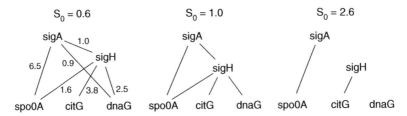

Figure A4. *Networks inferred using Mutual Information with three different significance levels S_0. With respect to the network in Figure A2 the network with $S_0 = 0.6 \sim 45\%$ has a structural difference equal to two being one edge a false positive. The network with $S_0 = 1.0 \sim 68\%$ has a structural difference of only one. Finally, the network with $S_0 = 2.6 \sim 99\%$ has a structural difference equal to four. The numbers next to the edges in the network with $S_0 = 0.6$ are the significance of the regulation, all being greater or equal than S_0. The missing edges in the same network were not selected because their significance was lower. Note that it is enough to calculate once the significance of all gene pairs. To construct a network with a certain S_0 just select all significances with $S \geq S_0$.*

---

[8] 79% of the Gaussian is covered within 1.23831 standard deviations.

To compare MI inference with the Bayesian technique we use the data in *Table A1* discretized with the threshold that gave the network with the lowest structural difference, $x_0 = 1.0$. In *Figure A4* we show three network structures obtained using three different significance levels and 100 000 surrogate data sets. Setting a high significance level generally causes one to miss many true interactions, and if the significance level is too low, false positives appear.

References

Abouheif, E. & Wray, G.A. (2002) Evolution of the gene network underlying wing polyphenism in ants. *Science* **297**: 249–252.

Affymetrix (2004) *Gene Chip Expression Analysis Technical Manual.* http://www.affymetrix.com

Albert, R. (2005) Scale-free networks in cell biology. *J. Cell Sci.* **118**: 4947–4957.

Albert, R. & Othmer, H.G. (2003) The topology of the regulatory interactions predicts the expression pattern of the segment polarity genes in *Drosophila melanogaster. J. Theor. Biol.* **223**: 1–18.

Alberts, B., Johnson, A., Lewis, J., Raff, M., Roberts, K. & Walter, P. (2002) *Molecular Biology of the Cell*, 4th Edn. Garland, New York.

Almaas, E. (2007) Biological impacts and context of network theory. *J. Exp. Biol.* **210**: 1548–1558.

Alvarez-Buylla, E.R., Benítez, M., Balleza Dávila, E., Chaos, A., Espinosa-Soto, C. & Padilla-Longoria, P. (2007) Gene regulatory network models for plant development. *Curr. Opin. Plant Biol.* **10**: 83–91.

Ambrose, B.A., Espinosa-Matías, S., Vázquez-Santana, S., Vergara-Silva, F., Martínez, E., Márquez-Guzmán, J. & Alvarez-Buylla, E.R. (2006) Comparative developmental series of the Mexican triurids support a euanthial interpretation for the unusual reproductive axes of *Lacandonia schismatica* (Pandanales, Liliopsida). *Am. J. Bot.* **93**: 15–35.

Aracena, J. & Demongeot, J. (2004) Mathematical methods for inferring regulatory networks interactions: Application to genetic regulation. *Acta. Biotheor.* **52**: 391–400.

Arenas, A., Diaz-Guilera, A. & Perez-Vicente, C.J. (2006) Synchronization reveals topological scales in complex networks. *Phys. Rev. Lett.* **96**: 114102.

Arthur, W. (2003) Developmental constraint and natural selection. *Evol. Dev.* **5**: 117–118.

Ball, C.A., Awad, I.A.B., Demeter, J. *et al.* (2005) The Stanford microarray database accommodates additional microarray platforms and data formats. *Nucleic Acids Res.* **33**: D580–D582.

Basso, K., Margolin, A.A., Stolovitzky, G., Klein, U., Dalla-Favera, R. & Califano, A. (2005) Reverse engineering of regulatory networks in human B cells. *Nature Genet.* **37(4)**: 382–390.

Beldade, P. & Brakefield, P.M. (2003) The difficulty of agreeing about constraints. *Evol. Dev.* **5**: 119–120.

Benítez, M., Espinosa-Soto, C., Padilla-Longoria, P., Díaz, J. & Alvarez-Buylla, E.R. (2007) Equivalent genetic regulatory networks in different contexts recover contrasting spatial cell patterns that resemble those in Arabidopsis root and leaf epidermis: a dynamic. *Int. J. Dev. Biol.* **51**: 139–155.

Berger, S.L. (2007) The complex language of chromatin regulation during transcription. *Nature* **447**: 407–412.

Blake, W.J., KÆrn, M., Cantor, C.R., & Collins, J.J. (2003) Noise in eukaryotic gene expression. *Nature* **422**: 633–637.

Chaos, A., Aldana, M., Espinosa-Soto, C., García Ponce de León, B., Garay, A. & Alvarez-Buylla, E.R. (2006) From genes to flower patterns and evolution: dynamic models of gene regulatory networks. *J. Plant Growth Regul.* **25**: 278–289.

Coen, E.S. & Meyerowitz, E.M. (1991) The war of the whorls: genetic interactions controlling flower development. *Nature* **353**: 31–37.

Damen, W.G.M. (2007) Evolutionary conservation and divergence of the segmentation process in arthropods. *Dev. Dyn.* **236**: 1379–1391.

Davidson, E.H. & Erwin, D.H. (2006) Gene regulatory networks and the evolution of animal body plans. *Science* **311**: 796–800.

de Visser, J.A.G.M., Hermisson, J., Wagner, G.P. *et al.* (2003) Perspective: Evolution and detection of genetic robustness. *Evolution* **57**: 1959–1972.

Diambra, L. & da Fontoura Costa, L. (2006) Pattern formation in a gene network model with boundary shape dependence. *Phys. Rev. E.* **73**: 031917.

Dolan, L. (2006) Positional information and mobile transcriptional regulators determine cell pattern in *Arabidopsis* root epidermis. *J. Exp. Bot.* **57**: 51–54.

Dumais, J. (2007) Can mechanics control pattern formation in plants? *Curr. Op. Plant Biol.* **10**: 58–62.

Elowitz, M.B., Levine A.J., Siggia, E.D. and Swain, P.S. (2002) Stochastic gene expression in a single cell. *Science* **297**: 1183–1186.

Espinosa-Soto, C., Padilla-Longoria, P. & Alvarez Buylla, E. (2004) A gene regulatory network model for cell-fate determination during *Arabidopsis thaliana* flower development that is robust and recovers experimental gene expression profiles. *Plant Cell* **16**: 2923–2939.

Fiering, S., Whitelaw, E. & Martin, D.I.K. (2000) To be or not to be active: the stochastic nature of enhancer action. *BioEssays* **22**: 381–387.

Friedman, N. (2004) Inferring cellular networks using probabilistic graphical models. *Science* **303**: 799–805.

Gershenson, C. (2002) *Classification of Random Boolean Networks*. arXiv:cs/0208001.

Gómez-Mena, C., de Folter, S., Costa, M.M.R., Angenent, G.C. & Sablowski, R. (2005) Transcriptional program controlled by the floral homeotic gene AGAMOUS during early organogenesis. *Development* **132**: 429–438.

Gonzalez, A.G., Naldi, A., Sánchez, L., Thieffry, D. & Chaouiya, C. (2006) GINsim: a software suite for the qualitative modelling, simulation and analysis of regulatory networks. *Biosystems* **84**: 91–100.

Heckerman, D. (1996) *A Tutorial On Learning with Bayesian Networks*. Technical Report No. MSR-TR-95-06, Microsoft Research, Redmond, Washington.

Henderson, I.R. & Jacobsen, S.E. (2007) Epigenetic inheritance in plants. *Nature* **447**: 418–424.

Hogeweg, P. (2000) Evolving mechanisms of morphogenesis: On the interplay between differential adhesion and cell differentiation. *J. Theor. Biol.* **203**: 317–333.

Huang, S. & Ingber, D.E. (2006) A non-genetic basis for cancer progression and metastasis: self-organizing attractors in cell regulatory networks. *Breast Dis.* **26**: 27–54.

Hume, D.A. (2000) Probability in transcriptional regulation and its implications for leukocyte differentiation and inducible gene expression. *Blood* **96**: 2323–2328.

Jackson, D., Leys, S.P., Hinman, V.F., Woods, R., Lavin, M.F. & Degnan, B.M. (2002) Ecological regulation of development: induction of marine invertebrate metamorphosis. *Int. J. Dev. Biol.* **46**: 679–86.

Jonsson, H., Heisler, M.G., Shapiro, B.E., Meyerowitz, E.M. & Mjolsness, E. (2006) An auxin-driven polarized transport model for phyllotaxis. *Proc. Natl Acad. Sci. USA* **103**: 1633–1638.

Jungnickel, D. (2002) *Graphs, Networks and Algorithms*. Springer, Berlin.

Kauffman, S. (1969) Metabolic stability and epigenesis in randomly constructed genetic nets. *J. Theor. Biol.* **22**: 437–467.

Klemm, K. & Bornholdt, S. (2005a) Stable and unstable attractors in Boolean networks. *Phys. Rev.* E **72**: 055101.

Klemm, K. & Bornholdt, S. (2005b) Topology of biological networks and reliability of information processing. *Proc. Natl Acad. Sci. USA* **102**: 18414–18419.

Lewin, B (2007) *Genes IX*, 9th Edn, Jones and Bartlett Publishers, Sudbury, MA.

Lewontin, R. (1974) *The Genetic Basis of Evolutionary Change*. Columbia University Press, New York.

Liang, S., Fuhrman, S. & Somogyi, R. (1998) REVEAL: A general reverse engineering algorithm for inference for genetic network architectures. In: *Pacific Symposium on Biocomputing 98* (eds R.B. Altmer, A.K. Dunker, L. Hunter and T.F. Klein). World Scientific, pp. 18–29.

Maynard Smith, J., Burian, R., Kauffman, S.A., Alberch, P., Campbell, J., Goodwin, B.C., Lande, R., Raup, D. & Wolpert, L. (1985) Developmental constraints and evolution. *Q. Rev. Biol.* **60**: 265–287.

Meinhardt, H. & Gierer, A. (2000) Pattern formation by local self-activation and lateral inhibition. *BioEssays* **22**: 753–760.

Meir, E., von Dassow, G., Munro, E. & Odell, G.M. (2002) Robustness, flexibility, and the role of lateral inhibition in the neurogenic network. *Curr. Biol.* **12**: 778–786.

Mendoza, L. & Alvarez-Buylla, E.R. (1998) Dynamics of the genetic regulatory network for *Arabidopsis thaliana* flower morphogenesis. *J. Theor. Biol.* **193**: 307–319.

Mendoza, L. & Alvarez-Buylla, E.R. (1999) Genetic control of flower morphogenesis in *Arabidopsis thaliana*: a logical analysis. *Bioinformatics* **15**: 593–606.

Mendoza, L. & Alvarez-Buylla, E.R. (2000) Genetic regulation of root hair development in *Arabidopsis thaliana*: a network model. *J. Theor. Biol.* **204**: 311–326.

Meyerowitz, E.M. (2002) Plants compared to animals: the broadest comparative study of development. *Science* **295**: 1482–1485.

Milo, R., Itzkovitz, S., Kashtan, N., Levitt, R., Shen-Orr, S., Ayzenshtat, I., Sheffer, M. & Alon, U. (2004) Superfamilies of evolved and designed networks. *Science* **303**: 1538–1542.

Neapolitan, R.E. (2004) *Learning Bayesian Networks*. Prentice Hall, Englewood Cliffs, NJ.

Ozbudak, E.M., Thattai, M., Kurtser, I., Grossman, A.D. & van Oudenaarden, A. (2002) Regulation of noise in the expression of a single gene. *Nat. Genet.* **31**: 69–73.

Paulsson, J. (2004) Summing up the noise in gene networks. *Nature* **427**: 415–418.

Pearl, J. (2000) *Causality*. Cambridge University Press, New York.

Perkins, T.J., Jaeger, J., Reinitz, J. & Glass, L. (2006) Reverse engineering the gap gene network of *Drosophila melanogaster*. *PloS Comp. Biol.* **2:** e51.

Pesch, M. & Hülskamp, M. (2004) Creating a two-dimensional pattern de novo during *Arabidopsis* trichome and root hair initiation. *Curr. Opin. Genet. Devel.* **14:** 422–427.

Plaza, R.G., Sánchez-Garduño, F., Padilla, P., Barrio, R.A. & Maini, P.K. (2005) The effect of growth and curvature on pattern formation. *J. Dyn. Diff. Eq.* **16:** 1093–1121.

Qi, J.P., Shao, S.H., Xie, J. & Zhu, Y. (2007) A mathematical model of P53 gene regulatory networks under radiotherapy. *BioSystems* **90:** 698-706.

Quackenbush, J. (2001) Computational genetics: Computational analysis of microarray data. *Nat. Rev. Gen.* **2:** 418–427.

Ribeiro, A.S. & Kauffman, S.A. (2007) Noisy attractors and ergodic sets in models of gene regulatory networks. *J. Theor. Biol.* **247:** 743-755.

Rice, S.H. (2000) The evolution of developmental interactions: epistasis, canalization and integration. In: *Epistasis and the Evolutionary Process* (eds J.B. Wolf, E.D. Brodie III and M.J. Wade). Oxford University Press, New York, pp. 82–98.

Ross, I.L., Browne, C.M. & Hume, D.A. (1994) Transcription of individual genes in eukaryotic cells occurs randomly and infrequently. *Immunol. Cell Biol.* **72:** 177–185.

Rossi, F.M., Kringstein, A.M., Spicher, A., Guicherit, O.M. & Blau, H.M. (2000) Transcriptional control: rheostat converted to on/off switch. *Mol. Cell* **6:** 723–728.

Rudall, P. (1987) *Anatomy of Flowering Plants: An Introduction to Structure and Development*. Edward Arnold, London.

Sánchez, L. & Thieffry, D. (2001) A logical analysis of the *Drosophila* gap gene system. *J. Theor. Biol.* **211:** 115–141.

Schlosser, G. & Wagner, G.P. (eds) (2004) *Modularity in Development and Evolution*. University of Chicago Press, Chicago, IL.

Segal, E., Shapira, M., Regev, A., Pe'er, D., Boststein, D., Koller, D. & Friedman, N. (2003) Module networks: identifying regulatory modules and their condition-specific regulators from gene expression data. *Nat. Genet.* **34(2):** 166–176.

Shmulevich, I., Dougherty, E.R., Kim, S. & Zhang, W. (2002) Probabilistic Boolean networks: a rule-based uncertainty model for gene regulatory networks. *Bioinformatics* **18:** 261–274.

Shmulevich, I., Lahdesmaki, H., Dougherty, E.R., Astola, J. & Zhang, W. (2003) The role of certain Post classes in Boolean network models of genetic networks. *Proc. Natl Acad. Sci. USA* **100:** 10734–10739.

Sick, S., Reinker, S., Timmer, J. & Schlake, T. (2006) WNT and DKK determine hair follicle spacing through a reaction-diffusion mechanism. *Science* **314:** 1447–1450.

Steuer, R., Kurths, J., Daub, C.O., Weise, J. & Selbig, J. (2002) The mutual information: Detecting and evaluating dependencies between variables. *Bioinformatics* **18 Suppl. 2:** S231–S240.

Templeton, A.R. (2000) Epistasis and complex traits. In *Epistasis and the Evolutionary Process* (eds J.B. Wolf, E.D. Brodie III and M.J. Wade). Oxford University Press, New York, pp. 41–57.

Thieffry, D. & Sánchez, L. (2004) Quantitative analysis of gene networks: Toward the delineation of cross-regulatory modules. In: *Modularity in Development and*

Evolution (eds G. Schlosser and G.P. Wagner). University of Chicago Press, Chicago, IL, pp. 222–243.

Thomas, R. (1991) Regulatory networks seen as asynchronous automata: A logical description. *J. Theor. Biol.* **153:** 1–23.

Tijsterman, M., Ketting, R.F. & Plasterk, R.H.A. (2002) The genetics of RNA silencing. *Ann. Rev. Gen.* **36:** 489–519.

Turing, A. (1952) The chemical basis of morphogenesis. *Philos. Trans. R. Soc. Lond. B. Biol. Sci.* **237:** 37–72.

Valente, A.X.C.N. & Cusick, M.E. (2006) Yeast protein interactome topology provides framework for coordinated-functionality. *Nucleic Acids Res.* **34:** 2812–2819.

Vergara-Silva, F., Espinosa S., Ambrose, A., Vázquez-Santana, S., Martínez-Mena A., Márquez-Guzmán, J., Martínez, E., Meyerowitz, E. & Alvarez-Buylla, E.R. (2003) Inside-out flower characteristic of *Lacandonia schismatica* evolved at least before its divergence from a closely related taxon, *Triuris brevistylis*. *Int. J. Plant Sciences* **164:** 345–357.

von Dassow, G. & Munro, E.M. (1999) Modularity in animal development and evolution: elements of a conceptual framework for EvoDevo. *J. Exp. Zool. (Mol. Dev. Evol.)* **285:** 307–325.

von Dassow, G. & Odell, G.M. (2002) Design and constraints of the *Drosophila* segment polarity module: robust spatial patterning emerges from intertwined cell state switches. *J. Exp. Zool. (Mol. Dev. Evol.)* **294:** 179–215.

von Dassow, G., Meir, E., Munro, E.M. & Odell, G.M. (2000) The segment polarity network is a robust developmental module. *Nature* **406:** 188–193.

Waddington, C.H. (1957) *The Strategy of the Genes. A Discussion of Some Aspects of Theoretical Biology*. MacMillan, New York.

Wagner, G.P. (2007) The developmental genetics of homology. *Nature* **8:** 473–479.

Wang, H., Ngwenyama, N., Liu, Y., Walker, J.C. & Zhang, S. (2007) Stomatal development and patterning are regulated by environmentally responsive mitogen-activated protein kinases in *Arabidopsis*. *Plant Cell* **19:** 63–73.

Wang, Y., Joshi, T., Zhang, X.-S., Xu, D. & Chen, L. (2006) Inferring gene regulatory networks from multiple microarray datasets. *Bioinformatics* **22:** 2413–2420.

Wuensche, A. (2001) *The DDLab Manual* and *Discrete Dynamics Lab* software. Online at www.santafe.edu/~wuensch/ddlab.html and www.ddlab.com

Yu, J., Smith, V., Wang, P., Hartemink, A. & Jarvis, E. (2002) Using Bayesian network inference algorithms to recover molecular genetic regulatory networks. In: *International Conference on Systems Biology 2002 (ICSB02)*, December, 2002.

Yu, J., Smith, V.A., Wang, P.P., Hartemink, A.J. & Jarvis, E.D. (2004) Advances to Bayesian network inference for generating causal networks from observational biological data. *Bioinformatics* **20(18):** 3594–3603.

7

Spatio-temporal dynamics of protein modification cascades

Boris N. Kholodenko and Herbert M. Sauro

In this chapter we wish to illustrate the use of an emerging systems biology approach to link molecular mechanisms to complex spatio-temporal dynamics of signal transduction cascades. We review recent work that combines quantitative measurements and computational modelling of signalling pathways to understand their spatio-temporal dynamics. Universal motifs commonly found in cellular signalling networks are protein-modification cycles, which are catalysed by opposing enzymes, such as a kinase and phosphatase. We demonstrate that simple cycles and cascades can generate complex temporal dynamics, including bistability and damped or sustained oscillations. The spatial separation of opposing reactions within protein-modification cycles results in the intracellular gradients of protein activities. These spatial gradients can guide pivotal physiological processes, such as cell motility and mitosis, but also impose a need for facilitated signal propagation, which involves trafficking of endosomes and signalling complexes along microtubules and travelling waves of phosphorylated kinases. Overall, we demonstrate the potential of systems biology to generate new insight and knowledge with quantitative and predictive explanatory power at the system level.

1 Introduction

Countless environmental cues received by plasma membrane receptors are processed and encoded into complex spatio-temporal response patterns of protein-modification networks and topological relocation of signalling proteins. There is no single protein or gene that is responsible for specificity of signalling by different receptors. Rather, signal specificity and cellular decisions are outputs of a complex interplay of multiple signalling processes that often involve feedback regulation by alterations in gene expression (Kholodenko, 2006; Shymko et al., 1997; Whitehead et al., 2000). This complex picture involves the temporal and spatial organization of signalling network activation. A subtle difference in receptor activation patterns and interaction kinetics may result in differential downstream responses and, eventually, in alterations in gene expression by signal-regulated transcription factors (Hoffmann et al., 2002; Marshall, 1995; Murphy et al., 2002, 2004; Werner et al., 2005).

For instance, the activation of Raf-1 (a direct downstream effector of Ras and the kinase for MEK, has been linked to such opposing responses as the induction of

DNA synthesis and growth inhibition (Lloyd *et al.*, 1997; Sewing *et al.*, 1997). A classical example is the distinct biological outcome of PC12 cell-line stimulation with epidermal growth factor (EGF) and nerve growth factor (NGF). EGF induces a transient activation of the extracellular signal regulated kinase (ERK), which results in proliferation, whereas a sustained ERK activation by NGF changes the cell fate and induces differentiation (Marshall, 1995). Likewise, sustained versus transient activation of the mitogen-activated protein kinase (MAPK) cascade was suggested to be a mechanism underlying receptor tyrosine kinase (RTK) specificity in EGF- and HGF-induced (hepatocyte growth factor) keratinocyte migration (McCawley *et al.*, 1999). However, the factors controlling the kinetics of MAPK cascades are intricate. MAPK cascades can generate bistable dynamics (where two stable 'on' and 'off' steady states coexist), abrupt switches, and oscillations (Altan-Bonnet and Germain, 2005; Bagowski *et al.*, 2003; Bhalla *et al.*, 2002; Blüthgen *et al.*, 2006; Kholodenko, 2000; Markevich *et al.*, 2004a; Wang *et al.*, 2006; Xiong and Ferrell, 2003), and their responses depend dramatically on their sub-cellular localization or recruitment to scaffolds (Harding *et al.*, 2005; Whitehurst *et al.*, 2004). This makes it difficult to comprehend the cellular response dynamics, using purely qualitative arguments.

This chapter illustrates the benefits of the application of systems analysis and modelling to the studies of protein-modification networks. We begin with a brief overview of modelling of growth factor signalling and challenges that face every mechanistic model. We describe reversible covalent modification cycles that form universal motifs of cellular signalling networks and show how cycles and cascades process and integrate signals and which feedback architecture enables robustness, linear or ultrasensitive responses, bistability and oscillations. Finally, we analyse the spatial dynamics of intracellular communication and show how gradients of protein activities are generated within a single cell. We propose that signalling endosomes, protein complexes assembled on scaffolds and travelling waves of protein modification spread signals over large intracellular distances.

2 Computational modelling of growth factor signalling

Different growth factors stimulate cell-surface receptors, which possess intrinsic tyrosine kinase activity, and are referred to as receptor tyrosine kinases (RTKs; De Meyts *et al.*, 2004; Schlessinger, 2000; Schlessinger and Ullrich, 1992). Ever since the EGF receptor (EGFR) was cloned two decades ago, RTKs have been in the limelight of scientific interest owing to their central role in the regulation of embryogenesis, cell survival, motility, proliferation, differentiation, glucose metabolism and apoptosis. EGFR, the cellular homologue of a retrovirally transduced oncogene (v-erbB), is highly expressed in many solid tumours, and this high expression level frequently correlates with poor prognosis (Mendelsohn and Baselga, 2003; Yarden and Sliwkowski, 2001). The cellular homologues of other oncogenes, such as Ras, Raf and Akt were shown to be intracellular effectors of EGFR, insulin receptor and other RTKs. Aberrant signalling by IR is implicated in type 2 diabetes, the frequency of which is increasing owing to changes in lifestyle, diet and environment (De Meyts *et al.*, 2004).

All RTKs consist of three major domains: An extracellular domain for ligand binding, a membrane-spanning segment and a cytoplasmic domain, which possesses tyrosine kinase activity and contains phosphorylation sites with tyrosine, serine and threonine residues. Following ligand binding, RTKs undergo dimerization (e.g. EGFR) or allosteric transitions (e.g. insulin receptor and insulin-like growth factor-1 receptor) that results in the activation of intrinsic tyrosine kinase activity. Autophosphorylation of RTKs transmits biochemical signals to cytoplasmic adaptor/target proteins, which contain characteristic protein domains, such as src homology (SH2 and SH3), phosphotyrosine binding (PTB) and pleckstrin homology (PH) domains, thereby triggering their mobilization to the cell surface (Kholodenko et al., 2000b; Pawson, 1995; Pawson et al., 2001; Schlessinger, 1994; Taniguchi et al., 2006; Van Obberghen et al., 2001). Subsequently, signals propagate through multiple interacting branches including the PLCγ, PI3K/Akt, JAK/STAT and MAPK pathways (see EGFR network diagrams in Kholodenko, 2006 and Oda et al., 2005).

Understanding of the temporal dynamics of cellular responses requires more than a list of binding partners and interactions. Since the 1990s, computational modelling has emerged as a novel tool to handle the rapidly growing information on the molecular parts list and the overwhelmingly complex interaction circuitry of signalling networks (Asthagiri and Lauffenburger, 2001; Bhalla and Iyengar, 1999; Bornheimer et al., 2004; Brightman and Fell, 2000; Csete and Doyle, 2002; Goldstein et al., 2004; Hatakeyama et al., 2003; Haugh et al., 2000; Heinrich et al., 2002; Kholodenko et al., 1999; Markevich et al., 2004b; Resat et al., 2003; Sasagawa et al., 2005; Schoeberl et al., 2002; Shvartsman et al., 2002; Tyson et al., 2001, 2003; Wolf and Arkin, 2003). Importantly, even the first mechanistic model of the EGFR network (Kholodenko et al., 1999) already included a combination of theoretical predictions and experimental validations, which now is deemed standard for systems biology (Citri and Yarden, 2006).

A critical challenge facing mechanistic modelling is the combinatorial increase in the number of emerging distinct species and states of the protein network being simulated (Faeder et al., 2003; Goldstein et al., 2004; Hlavacek et al., 2003; Morton-Firth and Bray, 1998). Many receptors and signalling proteins possess multiple docking sites, serving as scaffolds that generate a variety of heterogeneous multi-protein complexes, each involved in multiple parallel reactions (Figure 1A). Even initial steps in signal transduction can generate hundreds of thousands of distinct states (Goldstein et al., 2004), referred to as 'micro-states' of a network (Borisov et al., 2005, 2006). Several methods of handling this problem have been proposed, all based on specifying rules that automatically generate species and reactions. Programs implementing these methods include StochSim (Le Novere and Shimizu, 2001; Morton-Firth and Bray, 1998), BioNetGen (Blinov et al., 2004; Faeder et al., 2005), and Moleculizer (Lok and Brent, 2005). The entire micro-state network can either be generated in advance for deterministic simulations (Blinov et al., 2004; Faeder et al., 2005), or the species and reactions can be generated as needed during a stochastic simulation (Faeder et al., 2005; Lok and Brent, 2005; Morton-Firth and Bray, 1998).

An alternative 'domain-oriented' approach that approximates a mechanistic micro-state picture in terms of 'macro-variables', such as the phosphorylation levels and the fractions occupied by binding partners, was recently proposed (Borisov et al., 2005, 2006; Conzelmann et al., 2006). A necessary prerequisite for this macro-description is the presence of domains/sites that do not allosterically influence each other. Each macro-variable accounts for the states of a separate site on a protein and the domains that control this site. For instance, the state of each docking site on

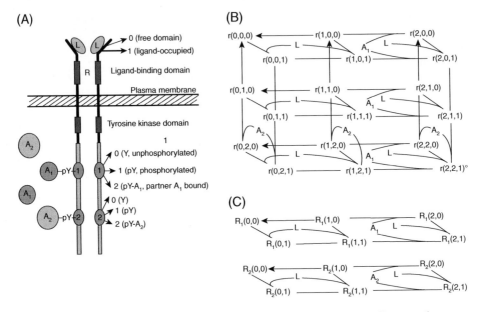

Figure 1. *Multiplicity of possible receptor states and reactions. (A) Digital flags 0 and 1 are assigned to the free (0) and ligand (L)-occupied (1) states of the ligand-binding domain of the receptor R. For each of two docking sites, numbers 0, 1 and 2, correspond to tyrosine residues that are unphosphorylated (Y), phosphorylated but unoccupied (pY), and phosphorylated and occupied by a binding partner (pY-A_i). (B) Fragment of the transition graph, where r(i,j;k) designate 18 states of the receptor R with the ligand-binding domain in state (k =0,1) and two docking sites in states (i =0,1,2) and (j =0,1,2). (C) Reduction of (B) in terms of macro-states. Each macro-variable, R₁(i,k) and R₂(j,k), describes the states of an individual docking site on R at a certain state k of the ligand-binding domain (referred to as a controlling site). The transition graph splits into two disconnected 6-state sub-graphs corresponding to single docking sites.*

a typical RTK is controlled by ligand binding and dimerization, and the macro-variable associated with that site may depend on the ligand and dimerization state of the receptor (illustrated in *Figures 1B* and *1C*). This domain-oriented framework drastically reduces the number of states and differential equations to be solved and, therefore, the computational cost of both deterministic and stochastic simulations. For instance, in a recently published model of the EGFR network, the use of a domain-oriented description reduces the number of equations and variables from hundreds of thousands to about 350 and thereby brings the analysis of the network within the realm of possibilities (Kiyatkin *et al.*, 2006).

3 Temporal dynamics of protein modification cascades

Cellular responses to changes in the external and internal environments include reversible covalent modification of proteins that already exist in a cell. These post-translational

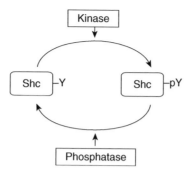

Figure 2. *Simple cycle of protein modification. This cycle is formed by a kinase and phosphatase, which phosphorylate and dephosphorylate a Tyr residue (Y) on a target protein (Shc).*

modifications dramatically change protein activities and underlie the physiological behaviour of all cells. In eukaryotes, post-translational protein modification reactions include phosphorylation of Tyr, Thr and Ser residues, ubiquitylation and sumoylation of Lys, methylation of Arg and Lys, acetylation of Lys, hydroxylation of Pro and other modifications (Seet *et al.*, 2006). Reversible covalent modification of proteins generates universal motifs of cell signalling: The cycle formed by two or more inter-convertible forms of a signalling protein, which is modified by two opposing enzymes (see *Figure 2*). Such opposing enzymes can be a kinase and phosphatase in the case of phosphorylation, ubiquitin ligase and deubiquitylating isopeptidase for ubiquityla-tion, methyltransferase and amine oxidase demethylase for methylation, and the cor-responding enzyme pairs for (de)acetylation and (de)hydroxylation. Another crucial example of a reversible modification cycle is the exchange of guanine nucleotides GDP and GTP on small G-proteins, such as proteins of the Ras family (*Figure 2*). A GTP-bound form of a small G-protein is usually active, whereas a GDP-bound form is inactive. The two opposing enzymes that catalyse the cycle are a guanine nucleotide exchange factor (GEF), such as SOS, and a GTP-hydrolysis activating protein (GAP).

A well-known property of protein modification cycles is their 'ultrasensitivity' to input signals, which occurs when the converting enzymes operate near saturation (Goldbeter and Koshland, 1981). Depending on the degree of saturation, the response of either interconvertible form ranges from a merely hyperbolic to a steep sigmoidal curve. It has been shown that sequestration of a signalling protein (which occurs when the concentration of its complexes with converting enzymes becomes compara-ble with its total concentration) significantly decreases sigmoidicity of responses (Blüthgen *et al.*, 2006). Likewise, ultrasensitivity can disappear if converting enzymes are inhibited or saturated by their products (Ortega *et al.*, 2002). Importantly, multi-site phosphorylation that occurs through a distributive, multi-collision mecha-nism was shown to increase the sensitivity dramatically, resisting the sequestration effect and leading to switch-like responses (Ferrell, 1997; Kholodenko *et al.*, 1998; Markevich *et al.*, 2004a; Ortega *et al.*, 2006; Salazar and Hofer, 2003).

3.1 *The role of feedback*

The notion of feedback is fundamental in biology. Almost any biological pathway has feedback. Feedback enables cells to monitor information about pathway activities in order to control the chemical processes along the pathways. Negative feedback regulation of a key enzyme by the end product of a metabolic pathway was discovered fifty years ago (Pardee and Yates, 1956; Umbarger, 1956). Since then feedback regulation has been described for metabolic, signalling and gene expression pathways. There are numerous examples of systems with feedback, which include both human-made and natural biological systems. The role of feedback was extensively studied in physics and engineering and the feedback function is now well understood (Sauro and Kholodenko, 2004).

As an illustration, we consider a linear chain of protein modification cycles, where an active form of an enzyme at each level (E_i^*) catalyses the transformation of an inactive form of the enzyme at the next level into its active form (E_{i+1}^*), *Figure 3*. The overall sensitivity (R) of the cascade to input changes is defined as the steady-state fractional change in the level of the activated target enzyme E_n^* divided by the fractional change in the signal S (that caused the change in the target),

$$R = d \log E_n^* / d \log S. \tag{1}$$

This essentially equals the percentage change in E_n^* brought about by a 1% change in S. If R is much greater than 1, the response curve (the steady-state concentration of E_n^* versus S) is steeply sigmoidal. If the response is approximated by the Hill equation, R can be related to the Hill coefficient.

Within a cascade the response of an individual level (i) to the immediately preceding level ($i-1$) can be quantified similarly to Equation (1), just as if there were a single level cascade,

$$r_i = d \log E_i^* / d \log E_{i-1}^* \tag{2}$$

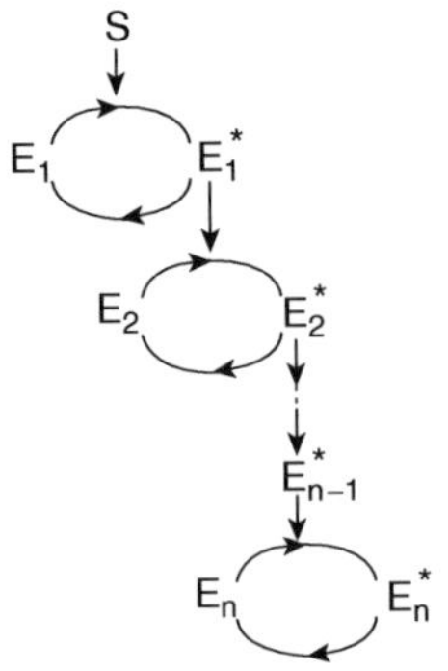

Figure 3. Linear pathway of protein modification cycles. Inactive and active enzyme forms are denoted as Ei and E_i^*, respectively. Positive feedback increases the sensitivity of the output enzyme (E_n^*) to the input signal (S), but may also lead to the existence of two stable steady states (bistability).

We will call r_i the local sensitivity of level i to the preceding level $i - 1$. From the chain rule of differentiation, it then follows that the overall sensitivity of a cascade with no feedback (R) is equal to the product of the local sensitivities at each level (Kholodenko et al., 1997). For example, for a cascade with three levels, such as the MAPK/ERK cascade, $R = r_1 \cdot r_2 \cdot r_3$. If the sensitivity at each level is more than 1, then merely having more levels can greatly increase the overall sensitivity of a cascade to the stimulus (Brown et al., 1997; Ferrell, 1997).

Negative feedback plus ultrasensitivity brings about oscillations in protein modification cascades. When the activated enzyme E_n^* at the bottom of a cascade immediately or indirectly affects some reaction(s) at the initial cascade level, this is referred to as a positive or negative feedback depending on whether E_n^* enhances or inhibits the activation level of the initial enzyme E_1^*. For instance, if the rate of the conversion ($v1$) of the inactive form E_1 in the active form E_1^* at the first cascade level is affected by E_n^*, the feedback strength F is given by,

$$F = d \log v_1 / d \log E_n^*. \tag{3}$$

Thus, the feedback strength is quantified as the percentage change in the reaction rate $v1$ brought about by a 1% change in E_n^*. In the case of activation, F is positive, and in the case of inhibition, F is negative. If a phosphorylation cascade is embedded into a feedback loop, the overall sensitivity (R_f) changes dramatically compared with the same cascade with no feedback (R) (Kholodenko et al., 1997; Sauro and Kholodenko, 2004),

$$R_F = R/(1 - F \cdot R). \tag{4}$$

From this equation, it follows that a negative feedback ($F < 0$) decreases the overall sensitivity of a cascade. Although negative feedback decreases the output to input amplification (the gain in engineering terms), it increases robustness of the system and stabilizes the gain (reviewed in Sauro and Kholodenko, 2004). In fact, it follows from Equation (4) that if $-F \cdot R$ exceeds 1, than the sensitivity R_F is determined mainly by the strength of feedback and depends less on the properties of individual levels within the feedback loop. However, if negative feedback is too strong, it can induce damped or sustained oscillations (Dibrov et al., 1982). Kholodenko (2000) showed that negative feedback can bring about sustained oscillations in a linear cascade of protein modification cycles, provided there is a sufficient degree of ultrasensitivity in individual cycles. In particular, the potential for oscillations in MAPK cascades was predicted (Kholodenko, 2000). *Figure 4* shows oscillations in the activity of doubly phosphorylated ERK for a simple model of the Ras/MAPK cascade.

Positive feedback leads to amplification and potentially to bistability and oscillations. In covalent modification cascades, positive feedback greatly increases the sensitivity of the target to the input signal (Kholodenko et al., 1997). In fact, Equation (4) illustrates that positive feedback ($F > 0$) brings about an increase in the sensitivity, as indeed was observed in *Xenopus* eggs (Ferrell and Machleder, 1998). In addition, positive feedback is the source of instabilities. Positive feedback can lead to bistability (Bhalla et al., 2002; Kholodenko, 2000; Xiong and Ferrell, 2003), but also, either alone

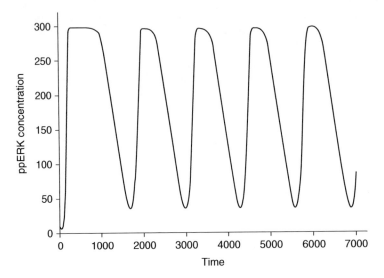

Figure 4. *Sustained oscillations in the ERK activity. The model (Kholodenko, 2000), in SBML format, was downloaded from the BioModel.net database (Le Novere, 2006; Model BIOMD0000000010) and simulated using the systems biology workbench (Sauro et al., 2003). The concentrations of active ERK (ppERK) are given in nM, and time is in seconds.*

or in combination with negative feedback, it can trigger oscillations (Goldbeter, 2002; Pomerening *et al.*, 2003; Sha *et al.*, 2003). From Equation (4) it follows that an increase in the cascade sensitivity becomes huge, if $F \cdot R$ approaches 1. The equality $F \cdot R = 1$ corresponds to the saddle-node bifurcation where two steady states, one unstable and one stable, are created or destroyed (Kholodenko, 2000). Importantly, the term $F \cdot R$ cannot become more than 1 and, therefore R_f could not be negative at any stable steady state.

We will illustrate the occurrence of bistability for a simple single-site protein modification cycle (for instance, the phosphorylation cycle) with positive feedback shown in *Figure 5A*. In the absence of feedback, a single-site cycle can generate only ultrasensitive, but not discontinuous, switches. Positive feedback from the phosphorylated form (T_p) of the target protein T to its kinase can render this cycle into a bistable switch. Assuming the kinase (v_{kin}) and phosphatase (v_{phos}) rates follow Michaelis–Menten kinetics, this system is described by a simple equation:

$$\frac{dT_p}{dt} = v_{kin} - v_{phos} = \frac{V_{max}^{kin} T}{(K_M^{kin} + T)} \cdot \frac{(1 + AT_p / K_A)}{(1 + T_p / K_A)} - \frac{V_{max}^{phosp} T_p}{(K_M^{phosp} + T_p)}, \quad T = T^{tot} - T_p \tag{5}$$

Here, V_{max}^{kin} and V_{max}^{phosp} are the maximal rates of the kinase and phosphatase, respectively, K_M^{kin} and K_M^{phosp} are the Michaelis constants, A and K_A are the kinetic constants of feedback activation, T^{tot} is the protein abundance. In a wide parameter range there are three distinct solutions to the steady-state relationship $v_{kin} = v_{phos}$. The low and high T_p

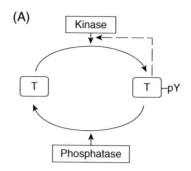

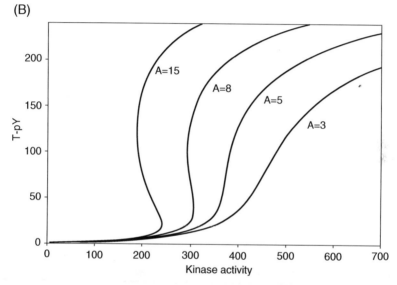

Figure 5. *Occurrence of bistability in a simple protein modification cycle with feedback. (A) Cycle of phosphorylation/dephosphorylation of the target protein (T) on a tyrosine residue (Y). The phosphorylated form T-pY of the protein T activates its own kinase. (B) One-dimensional bifurcation diagram shows the dependences of steady-state concentrations of the phosphorylated form T-pY on the kinase activity at different strength of positive feedback (parameter A). The kinase activity is given in nM/s; the values of the parameter A are shown on the curves; the other parameter values are* $V_{max}^{phos} = 200\ nM/s$; $K_M^{kin} = 500\ nM$; $K_M^{phosp} = 10\ nM$; $KA = 500\ nM$; $T^{tot} = 300\ nM$.

concentrations correspond to stable 'Off' and 'On' states, whereas the intermediate state is unstable. In the bistability range, the steady-state dependence of T_p on the input kinase (phosphatase) activity (known as one-dimensional bifurcation diagram) displays hysteresis, the hallmark of bistability. Figure 5B shows a family of such dependences of steady-state concentrations of the phosphorylated form on the kinase activity at different strength of positive feedback (parameter A). We see that increasing the strength of positive feedback leads to an increase in the sigmoidicity and eventually to hysteresis.

4 Spatial gradients of protein activities within a cell

Often, two opposing enzymes of a protein modification cycle are spatially separated within a cell. For instance, a kinase can be localized to a scaffold, supra-molecular structure or a cellular membrane, whereas a phosphatase can be homogeneously distributed in the surrounding area of the cytoplasm. In this case, the spatial gradient of a target cytoplasmic protein can occur, with the high level of phosphorylation of this protein in the close vicinity of the scaffold or the membrane and the low phosphorylation level at distant cytoplasmic areas (Brown and Kholodenko, 1999; Kholodenko *et al.*, 2000a). Likewise, the spatial gradient of a GTP-bound form of a small G-protein can occur if the GEF for that protein is confined to a supra-molecular structure, whereas a GAP freely diffuses in the cytoplasm (Carazo-Salas *et al.*, 1999). Given measured values of protein diffusivity and kinase and phosphatase activities, it was estimated that phosphoprotein gradients can be large within the intracellular space (Brown and Kholodenko, 1999; Kholodenko *et al.*, 2000a). We illustrate these findings by analysing a phosphorylation/dephosphorylation cycle where a kinase is confined to the plasma membrane, whereas the opposing phosphatase and interconvertable protein are diffusible molecules. The spatio-temporal dynamics of the phosphorylated form c_p of the interconvertible protein is described by the following equation, referred to as the reaction-diffusion equation (Brown and Kholodenko, 1999):

$$\frac{\partial c_p}{\partial t} = D\Delta c_p - v_p(c_p). \tag{6}$$

Here D is the protein diffusion coefficient in the cytosol (assumed to be equal for the phosphorylated and unphosphorylated forms) and v_p is the phosphatase rate. Assuming that the phosphatase is far from saturation, $v_p = k_p c_p$ ($k_p = V_{max}/K_m$ is the observed first-order rate constant), and considering a spherical cell, this equation simplifies as follows:

$$\frac{\partial c_p(r,t)}{\partial t} = \frac{D}{r^2}\frac{\partial}{\partial r}\left(r^2 \frac{c_p}{\partial r}\right) - k_p c_p. \tag{7}$$

At the cell membrane ($r = L$, where L is the cell radius), the rate of protein phosphorylation by the kinase (v_{kn}^{mem}) should be equal to the rate of phosphoprotein diffusion into the cell interior; at the cell centre ($r = 0$) there should be no diffusion flux. This determines the two boundary conditions:

$$D\frac{\partial c_p}{\partial r}\bigg|_{r=L} = v_{kn}^{mem}, \qquad D\frac{\partial c_p}{\partial r}\bigg|_{r=L} = 0. \tag{8}$$

The steady-state solution to Equation (7) with the boundary conditions (8) reads (Brown and Kholodenko, 1999):

$$c_p(r) = \frac{const \cdot (\exp(\alpha r) - \exp(-\alpha r))}{r}; \quad \alpha = \sqrt{k_p/D}. \tag{9}$$

The constant multiplier in Equation (9) depends on the rate of the kinase reaction. As a consequence, the relative difference in concentration of the phosphorylated form between the cell membrane and the centre of the cell is determined only by the phosphatase activity k_p, the protein diffusivity D and the cell size L:

$$\frac{c_p(L)-c_p(0)}{c_p(0)} = \frac{\exp(\alpha L)-\exp(-\alpha L)}{2\alpha L} - 1; \quad \alpha = \sqrt{k_p/D}. \tag{10}$$

We can now estimate the relative gradient using Equation (10). Eukaryotic cell radii vary from 5 to 50 microns; we assume a radius of 10 µm, which is the typical hepatocyte radius. The diffusion coefficient of a soluble protein in the cytosol is estimated to be of the order of 1–10 µm² s⁻¹ (although it can be considerably lower if the protein reversibly binds to immobile components of the cell), and values for k_p range from about 0.1 to 10 s⁻¹ (reviewed in Kholodenko, 2003). We will use a D value of 5 µm² s⁻¹ here for illustrative purposes. With k_p values of 0.1, 1, 10 and 100 s⁻¹ the relative concentration gradients for the phosphorylated form, calculated from Equation (10), are 0.4, 9, 50 000, and 3×10^{17}, respectively. This means that for $k_p = 1$, there is an approximately 9-fold lower concentration of the phosphorylated form of the protein in the centre of the cell than at the cell membrane, that is, the gradient is large. And for $k_p \gg 1$ there can be no detectable level of the phosphorylated form at the centre of the cell.

This estimate of the spatial gradients of protein activities that would occur within a single cell was one of the success stories of network modelling. These theoretical predictions have materialized recently, when fluorescence resonance energy transfer-based biosensors enabled discoveries of intracellular gradients of the active form of the small GTPase Ran (Kalab et al., 2002) and the phosphorylated form of stathmin-oncoprotein 18 (Op18/stathmin) involved in the regulation of the microtubule polymerization (Caudron et al., 2005).

Cascades facilitate the propagation of protein modification signals. Our calculations above suggest that the spatial propagation of protein phosphorylation that is initiated at the plasma membrane can be terminated by phosphatases even before the phosphorylated form reaches the nucleus. Kinase cascades that start at the membrane will result in overlapping gradients of the phosphorylated forms of kinases that reach further into the cell. We have recently shown that the phosphorylated kinases at descending levels of the cascade reach further into the cell; this may be one reason that cascades exist (Kholodenko, 2006). Passing a message from the cell membrane to the nucleus requires a cascade of kinases and not a single cycle.

4.1 *Waves of protein phosphorylation*

Although the existence of more levels in a cascade facilitates signal transfer, robust signal transduction over long distances may require additional means, such as vesicular or non-vesicular transport of phosphorylated kinases (Kholodenko, 2002; Miaczynska et al., 2004; Perlson et al., 2005) and travelling waves of protein phosphorylation (Kholodenko, 2003; Reynolds et al., 2003; Tischer and Bastiaens, 2003). Molecular motor-mediated movement of the endosomes and kinase complexes along microtubules is remarkably distinct from chaotic diffusive motion, and it is able to prevent the formation of precipitous reaction-diffusion gradients (Howe, 2005; Howe and Mobley, 2004; Kholodenko, 2002).

Chemical waves are well documented in many facets of cell life. For example: calcium waves, involved in intracellular signalling and intercellular communications (Berridge, 1997; Berridge et al., 2003), membrane polarization waves in nerve axons and myocardial fibres (Keener and Sneyd, 1998; Scott, 1975), travelling metabolic waves of NAD(P)H within neutrophils (Kindzelskii and Petty, 2002), aggregation of Dictyostelium amoebas and autocrine-mediated intra- and intercellular waves (Maly et al., 2004; Pribyl et al., 2003; Shvartsman et al., 2002). A key feature of propagating protein phosphorylation waves is that a particular molecule of phosphoprotein would not have to diffuse all the way from the region where the protein is phosphorylated to the distant target area. Rather, the spread in the space emerges as a result of a combination of diffusion and rapid phosphorylation. As a necessary prerequisite of a phosphorelay wave, cytoplasmic reactions should be capable of elevating the phosphorylation level at any given spatial location following the arrival of phosphoprotein molecules by diffusion. This local amplification may arise from a positive feedback enabling a reaction cascade to display switch-like, bistable behaviour (Kholodenko, 2006). In fact, we recently demonstrated that bistability in covalent modification cascades can result in novel types of phosphoprotein waves that relay extracellular signals from the plasma membrane to distant targets (Markevich et al., 2006). Kinases stimulated on one side of the cell (e.g. in development, sensing of extracellular gradients, or as a result of non-uniform attachment to other cells or substrata) will propagate asymmetric distribution of phosphoproteins and dependent activities through bistable phosphorylation cascades.

5 Concluding remarks

In this short chapter we have been able to describe a few of the dynamical properties of simple signalling motifs found in eukaryotic cells. In particular we have discussed the significant influence of negative and positive feedback loops on temporal behaviour of both single- and multiple-layered phosphorylation cycles. Even in the simple systems discussed in this chapter one can observe a multitude of interesting behaviour, including bistability, oscillations, robustness, frequency filtration. Undoubtedly, other behaviours have yet to be uncovered. Two significant issues remain however, these include the consequences of noise on system dynamics and the generation of appropriate experimental data to support or refute our understanding of these systems.

Noise, as a result of molecular collisions, is an inherent property of biochemical networks. In many cases, the existence of noise can be ignored because the number of particles taking part in a particular process is large compared to the noise itself. In other cases where the number of molecules is very low, the effect of noise can be significant (Arkin et al., 1998). The importance of noise in biochemical networks has only recently being a focus of research, largely as a result of new measurement techniques. Prokaryotic systems especially have been shown to exhibit noisy dynamics (Elowitz et al., 2002). Given the existence of noise in these systems one may speculate whether noise impedes the ability to transmit coherent signals and would therefore be suppressed through specific regulatory mechanisms (Becskei and Serrano, 2000; Dublanche et al., 2006). One may also speculate whether cells exploit noise in some way, in some cases this appears to be the case (Arkin *et al.*, 1998). These questions remain largely unanswered; though research is beginning to highlight the importance of noise. Fortunately, signalling proteins, at least in eukaryotic cells, are present in relatively high numbers (10 000 to 100 000 copies) although receptor molecules are present in much small numbers. One may therefore suspect that the effects of noise will be manifest at signal reception and the transmission of the signal to the immediate downstream targets. The recent work by Altan-Bonnet and Germain (2005) provides some very interesting insights in the mechanisms that cells have evolved to robustly detect very small concentration changes and how networks have evolved to operate at different scales.

Without doubt the greatest impediment to improving our understanding of how signalling networks operate is the dearth of suitable data to test and refine our models. Unlike systematic systems biology where large amounts of data on many variables is collected to discern patterns of behaviour, validating dynamical models requires the collection of high-resolution temporal data on specific variables of interest often with specific hypotheses in mind. Current methodologies for monitoring cellular protein states include Western blots and less specific imaging-based techniques such as fluorescent GTP proteins or more recent tools such as FlAsH and ReAsH (Miyawaki *et al.*, 2003). Many of these techniques have drawbacks. Western blots, though highly specific, are expensive and tend to have low temporal resolution; fluorescent markers, particularly fluorescent proteins, on the other hand, are unable to discern specific phosphorylated states, but can be used to generate data at high temporal resolution, often in real time and *in vivo*. Of potential interest to dynamical systems biology is the emerging field of aptamer research (Isaacs *et al.*, 2006), which combines the specificity of Western blots with the high resolution capability of *in vivo* fluorescent probes. Aptamers are RNA molecules that can bind specific molecules and activate, via a conformational change, a variety of downstream reporters including fluorescent probes. Aptamers are currently better at recognizing and binding small molecules but such technologies might permit researchers in the future to monitor metabolite or specific protein concentrations in real-time. The generation of high-resolution spatial and temporal data series will enable exciting theoretical developments to lead to novel biological insights.

References

Altan-Bonnet, G. & Germain, R.N. (2005) Modeling T cell antigen discrimination based on feedback control of digital ERK responses. *PLoS Biol.* **3**: e356.

Arkin, A., Ross, J. & McAdams, H.H. (1998) Stochastic kinetic analysis of developmental pathway bifurcation in phage lambda-infected *Escherichia coli* cells. *Genetics* **149**: 1633–1648.

Asthagiri, A.R. & Lauffenburger, D.A. (2001) A computational study of feedback effects on signal dynamics in a mitogen-activated protein kinase (mapk) pathway model. *Biotechnol. Prog.* **17**: 227–239.

Bagowski, C.P., Besser, J., Frey, C.R. & Ferrell, J.E. (2003) The JNK cascade as a biochemical switch in mammalian cells. Ultrasensitive and all-or-none responses. *Curr. Biol.* **13**: 315–320.

Becskei, A. & Serrano, L. (2000) Engineering stability in gene networks by autoregulation. *Nature* **405**: 590–593.

Berridge, M.J. (1997) Elementary and global aspects of calcium signalling. *J. Physiol.* **499(2):** 291–306.

Berridge, M.J., Bootman, M.D. & Roderick, H.L. (2003) Calcium signalling: dynamics, homeostasis and remodelling. *Nat. Rev. Mol. Cell Biol.* **4:** 517–529.

Bhalla, U.S. & Iyengar, R. (1999) Emergent properties of networks of biological signaling pathways. *Science* **283:** 381–387.

Bhalla, U.S., Ram, P.T. & Iyengar, R. (2002) MAP kinase phosphatase as a locus of flexibility in a mitogen-activated protein kinase signaling network. *Science* **297:** 1018–1023.

Blinov, M.L., Faeder, J.R., Goldstein, B. & Hlavacek, W.S. (2004) BioNetGen: software for rule-based modeling of signal transduction based on the interactions of molecular domains. *Bioinformatics.* **20:** 3289-3291.

Blüthgen, N., Bruggeman, F.J., Legewie, S., Herzel, H., Westerhoff, H.V. & Kholodenko, B.N. (2006) Effects of sequestration on signal transduction cascades. *Febs. J.* **273:** 895–906.

Borisov, N.M., Markevich, N.I., Hoek, J.B. & Kholodenko, B.N. (2005) Signaling through receptors and scaffolds: independent interactions reduce combinatorial complexity. *Biophys. J.* **89:** 951–966.

Borisov, N.M., Markevich, N.I., Hoek, J.B. & Kholodenko, B.N. (2006) Trading the micro-world of combinatorial complexity for the macro-world of protein interaction domains. *Biosystems* **83:** 152–166.

Bornheimer, S.J., Maurya, M.R., Farquhar, M.G. & Subramaniam, S. (2004) Computational modeling reveals how interplay between components of a GTPase-cycle module regulates signal transduction. *Proc. Natl Acad. Sci. USA* **101:** 15899–15904.

Brightman, F.A. & Fell, D.A. (2000) Differential feedback regulation of the MAPK cascade underlies the quantitative differences in EGF and NGF signalling in PC12 cells. *FEBS Lett.* **482:** 169–174.

Brown, G.C. & Kholodenko, B.N. (1999) Spatial gradients of cellular phospho-proteins. *FEBS Lett.* **457:** 452–454.

Brown, G.C., Hoek, J.B. & Kholodenko, B.N. (1997) Why do protein kinase cascades have more than one level? *Trends Biochem. Sci.* **22:** 288.

Carazo-Salas, R.E., Guarguaglini, G., Gruss, O.J., Segref, A., Karsenti, E. & Mattaj, I.W. (1999) Generation of GTP-bound Ran by RCC1 is required for chromatin-induced mitotic spindle formation. *Nature* **400:** 178–181.

Caudron, M., Bunt, G., Bastiaens, P. & Karsenti, E. (2005) Spatial coordination of spindle assembly by chromosome-mediated signaling gradients. *Science* **309:** 1373–1376.

Citri, A. & Yarden, Y. (2006) EGF-ERBB signalling: towards the systems level. *Nat. Rev. Mol. Cell Biol.* **7:** 505–516.

Conzelmann, H., Saez-Rodriguez, J., Sauter, T., Kholodenko, B.N. & Gilles, E.D. (2006) A domain-oriented approach to the reduction of combinatorial complexity in signal transduction networks. *BMC Bioinformatics* **7:** 34 doi:10.1186/1471-2105-1187-1134.

Csete, M.E. & Doyle, J.C. (2002) Reverse engineering of biological complexity. *Science* **295:** 1664–1669.

De Meyts, P., Palsgaard, J., Sajid, W., Theede, A.M. & Aladdin, H. (2004) Structural biology of insulin and IGF-1 receptors. *Novartis Found Symp.* **262:** 160–171; discussion 171–176, 265–268.

Dibrov, B.F., Zhabotinsky, A.M. & Kholodenko, B.N. (1982) Dynamic stability of steady states and static stabilization in unbranched metabolic pathways. *J. Math. Biol.* **15:** 51–63.

Dublanche, Y., Michalodimitrakis, K., Kummerer, N., Foglierini, M. & Serrano, L. (2006) Noise in transcription negative feedback loops: simulation and experimental analysis. *Mol. Syst. Biol.* **2:** 41.

Elowitz, M.B., Levine, A.J., Siggia, E.D. & Swain, P.S. (2002) Stochastic gene expression in a single cell. *Science* **297:** 1183–1186.

Faeder, J.R., Hlavacek, W.S., Reischl, I., Blinov, M.L., Metzger, H., Redondo, A., Wofsy, C. & Goldstein, B. (2003) Investigation of early events in Fc epsilon RI-mediated signaling using a detailed mathematical model. *J. Immunol.* **170:** 3769–3781.

Faeder, J.R., Blinov, M.L., Goldstein, B. & Hlavacek, W.S. (2005) Rule-based modeling of biochemical networks. *Complexity* **10:** 22–41.

Ferrell, J.E., Jr. (1997) How responses get more switch-like as you move down a protein kinase cascade. *Trends Biochem. Sci.* **22:** 288–289.

Ferrell, J.E., Jr. & Machleder, E.M. (1998) The biochemical basis of an all-or-none cell fate switch in *Xenopus* oocytes. *Science* **280:** 895–898.

Goldbeter, A. (2002) Computational approaches to cellular rhythms. *Nature* **420:** 238–245.

Goldbeter, A. & Koshland, D.E., Jr. (1981) An amplified sensitivity arising from covalent modification in biological systems. *Proc. Natl Acad. Sci. USA* **78:** 6840–6844.

Goldstein, B., Faeder, J.R. & Hlavacek, W.S. (2004) Mathematical and computational models of immune-receptor signalling. *Nat. Rev. Immunol.* **4:** 445–456.

Harding, A., Tian, T., Westbury, E., Frische, E. & Hancock, J.F. (2005) Subcellular localization determines MAP kinase signal output. *Curr. Biol.* **15:** 869–873.

Hatakeyama, M., Kimura, S., Naka, T. *et al.* (2003) A computational model on the modulation of mitogen-activated protein kinase (MAPK) and Akt pathways in heregulin-induced ErbB signalling. *Biochem. J.* **373:** 451–463.

Haugh, J.M., Wells, A. & Lauffenburger, D.A. (2000) Mathematical modeling of epidermal growth factor receptor signaling through the phospholipase C pathway: mechanistic insights and predictions for molecular interventions. *Biotechnol. Bioeng.* **70:** 225–238.

Heinrich, R., Neel, B.G. & Rapoport, T.A. (2002) Mathematical models of protein kinase signal transduction. *Mol. Cell* **9:** 957–970.

Hlavacek, W.S., Faeder, J.R., Blinov, M.L., Perelson, A.S. & Goldstein, B. (2003) The complexity of complexes in signal transduction. *Biotechnol. Bioeng.* **84:** 783–794.

Hoffmann, A., Levchenko, A., Scott, M.L. & Baltimore, D. (2002) The Ikappa B-NF-kappaB signaling module: temporal control and selective gene activation. *Science* **298:** 1241–1245.

Howe, C.L. (2005) Modeling the signaling endosome hypothesis: why a drive to the nucleus is better than a (random) walk. *Theor. Biol. Med. Model* **2:** 43.

Howe, C.L. & Mobley, W.C. (2004) Signaling endosome hypothesis: A cellular mechanism for long distance communication. *J. Neurobiol.* **58**: 207–216.

Isaacs, F.J., Dwyer, D.J. & Collins, J.J. (2006) RNA synthetic biology. *Nat. Biotechnol.* **24**: 545–554.

Kalab, P., Weis, K. & Heald, R. (2002) Visualization of a Ran-GTP gradient in interphase and mitotic *Xenopus* egg extracts. *Science* **295**: 2452–2456.

Keener, J. & Sneyd, J. (1998) *Mathematical Physiology.* Springer, New York.

Kholodenko, B.N. (2000) Negative feedback and ultrasensitivity can bring about oscillations in the mitogen-activated protein kinase cascades. *Eur. J. Biochem.* **267**: 1583–1588.

Kholodenko, B.N. (2002) MAP kinase cascade signaling and endocytic trafficking: a marriage of convenience? *Trends Cell Biol.* **12**: 173–177.

Kholodenko, B.N. (2003) Four-dimensional organization of protein kinase signaling cascades: the roles of diffusion, endocytosis and molecular motors. *J. Exp. Biol.* **206**: 2073–2082.

Kholodenko, B.N. (2006) Cell-signalling dynamics in time and space. *Nat. Rev. Mol. Cell Biol.* **7**: 165–176.

Kholodenko, B.N., Hoek, J.B., Westerhoff, H.V. & Brown, G.C. (1997) Quantification of information transfer via cellular signal transduction pathways. *FEBS Lett.* **414**: 430–434.

Kholodenko, B.N., Hoek, J.B., Brown, G.C. & Westerhoff, H.V. (1998) Control analysis of cellular signal transduction pathways. In: *BioThermoKinetics in the Post Genomic Era* (eds C., Larsson, I. L., Pahlman, and L. Gustafsson,). Göteborg, pp. 102–107.

Kholodenko, B.N., Demin, O.V., Moehren, G. & Hoek, J.B. (1999) Quantification of short term signaling by the epidermal growth factor receptor. *J. Biol. Chem.* **274**: 30169–30181.

Kholodenko, B.N., Brown, G.C. & Hoek, J.B. (2000a) Diffusion control of protein phosphorylation in signal transduction pathways. *Biochem. J.* **350(3)**: 901–907.

Kholodenko, B.N., Hoek, J.B. &Westerhoff, H.V. (2000b) Why cytoplasmic signalling proteins should be recruited to cell membranes. *Trends Cell Biol.* **10**: 173–178.

Kindzelskii, A.L. & Petty, H.R. (2002) Apparent role of traveling metabolic waves in oxidant release by living neutrophils. *Proc. Natl Acad. Sci. USA* **99**: 9207–9212.

Kiyatkin, A., Aksamitiene, E., Markevich, N.I., Borisov, N.M., Hoek, J.B. & Kholodenko, B.N. (2006) Scaffolding protein Grb2-associated binder 1 sustains epidermal growth factor-induced mitogenic and survival signaling by multiple positive feedback loops. *J. Biol. Chem.* **281**: 19925–19938.

Le Novere, N. (2006) Model storage, exchange and integration. *BMC Neurosci.* **7 Suppl. 1:** S11.

Le Novere, N. & Shimizu, T.S. (2001) STOCHSIM: modelling of stochastic biomolecular processes. *Bioinformatics* **17**: 575–576.

Lloyd, A.C., Obermuller, F., Staddon, S., Barth, C.F., McMahon, M. & Land, H. (1997) Cooperating oncogenes converge to regulate cyclin/cdk complexes. *Genes Dev.* **11**: 663–677.

Lok, L. & Brent, R. (2005) Automatic generation of cellular reaction networks with Moleculizer 1.0. *Nat. Biotechnol.* **23**: 131–136.

Maly, I.V., Wiley, H.S. & Lauffenburger, D.A. (2004) Self-organization of polarized cell signaling via autocrine circuits: computational model analysis. *Biophys. J.* **86**: 10–22.

Markevich, N.I., Hoek, J.B. & Kholodenko, B.N. (2004a) Signaling switches and bistability arising from multisite phosphorylation in protein kinase cascades. *J. Cell Biol.* **164**: 353–359.

Markevich, N.I., Moehren, G., Demin, O., Kiyatkin, A., Hoek, J.B. & Kholodenko, B.N. (2004b) Signal processing at the Ras circuit: What shapes Ras activation patterns? *IEE Syst. Biol.* **1**: 104–113.

Markevich, N.I., Tsyganov, M.A., Hoek, J.B. & Kholodenko, B.N. (2006) Long-range signaling by phosphoprotein waves arising from bistability in protein kinase cascades. *Mol. Syst. Biol.* **2**: 61.

Marshall, C.J. (1995) Specificity of receptor tyrosine kinase signaling: transient versus sustained extracellular signal-regulated kinase activation. *Cell* **80**: 179–185.

McCawley, L.J., Li, S., Wattenberg, E.V. & Hudson, L.G. (1999) Sustained activation of the mitogen-activated protein kinase pathway. A mechanism underlying receptor tyrosine kinase specificity for matrix metalloproteinase-9 induction and cell migration. *J. Biol. Chem.* **274**: 4347–4353.

Mendelsohn, J. & Baselga, J. (2003) Status of epidermal growth factor receptor antagonists in the biology and treatment of cancer. *J. Clin. Oncol.* **21**: 2787–2799.

Miaczynska, M., Pelkmans, L. & Zerial, M. (2004) Not just a sink: endosomes in control of signal transduction. *Curr. Opin. Cell Biol.* **16**: 400–406.

Miyawaki, A., Sawano, A. & Kogure, T. (2003) Lighting up cells: labelling proteins with fluorophores. *Nat. Cell Biol.* **Suppl.**: S1–7.

Morton-Firth, C.J. & Bray, D. (1998) Predicting temporal fluctuations in an intracellular signalling pathway. *J. Theor. Biol.* **192**: 117–128.

Murphy, L.O., Smith, S., Chen, R.H., Fingar, D.C. & Blenis, J. (2002) Molecular interpretation of ERK signal duration by immediate early gene products. *Nat. Cell Biol.* **4**: 556–564.

Murphy, L.O., MacKeigan, J.P. & Blenis, J. (2004) A network of immediate early gene products propagates subtle differences in mitogen-activated protein kinase signal amplitude and duration. *Mol. Cell Biol.* **24**: 144–153.

Oda, K., Matsuoka, Y., Funahashi, A. & Kitano, H. (2005) A comprehensive pathway map of epidermal growth factor receptor signaling. *Mol. Syst. Biol.* doi: 10.1038/msb4100014.

Ortega, F., Acerenza, L., Westerhoff, H.V., Mas, F. & Cascante, M. (2002) Product dependence and bifunctionality compromise the ultrasensitivity of signal transduction cascades. *Proc. Natl Acad. Sci. USA* **99**: 1170–1175.

Ortega, F., Garces, J.L., Mas, F., Kholodenko, B.N. & Cascante, M. (2006) Bistability from double phosphorylation in signal transduction. *Febs. J.* **273**: 3915–3926.

Pardee, A.B. & Yates, R.A. (1956) Control of pyrimidine biosynthesis in *Escherichia coli* by a feed-back mechanism. *J. Biol. Chem.* **221**: 757–770.

Pawson, T. (1995) Protein modules and signalling networks. *Nature* **373**: 573–580.

Pawson, T., Gish, G.D. & Nash, P. (2001) SH2 domains, interaction modules and cellular wiring. *Trends Cell Biol.* **11**: 504–511.

Perlson, E., Hanz, S., Ben-Yaakov, K., Segal-Ruder, Y., Seger, R. & Fainzilber, M. (2005) Vimentin-dependent spatial translocation of an activated MAP kinase in injured nerve. *Neuron* **45**: 715–726.

Pomerening, J.R., Sontag, E.D. & Ferrell, J.E. (2003) Building a cell cycle oscillator: hysteresis and bistability in the activation of Cdc2. *Nat. Cell Biol.* **5**: 346–351.

Pribyl, M., Muratov, C.B. & Shvartsman, S.Y. (2003) Long-range signal transmission in autocrine relays. *Biophys. J.* **84**: 883–896.

Resat, H., Ewald, J.A., Dixon, D.A. & Wiley, H.S. (2003) An integrated model of epidermal growth factor receptor trafficking and signal transduction. *Biophys. J.* **85**: 730–743.

Reynolds, A.R., Tischer, C., Verveer, P.J., Rocks, O. & Bastiaens, P.I. (2003) EGFR activation coupled to inhibition of tyrosine phosphatases causes lateral signal propagation. *Nat. Cell Biol.* **5**: 447–453.

Salazar, C. & Hofer, T. (2003) Allosteric regulation of the transcription factor NFAT1 by multiple phosphorylation sites: a mathematical analysis. *J. Mol. Biol.* **327**: 31–45.

Sasagawa, S., Ozaki, Y., Fujita, K. & Kuroda, S. (2005) Prediction and validation of the distinct dynamics of transient and sustained ERK activation. *Nat. Cell Biol.* **7**: 365–373.

Sauro, H.M., Hucka, M., Finney, A., Wellock, C., Bolouri, H., Doyle, J. & Kitano, H. (2003) Next generation simulation tools: the Systems Biology Workbench and BioSPICE integration. *Omics* **7**: 355–372.

Sauro, H.M. & Kholodenko, B.N. (2004) Quantitative analysis of signaling networks. *Prog. Biophys. Mol. Biol* **86**: 5–43.

Schlessinger, J. (1994) SH2/SH3 signaling proteins. *Curr. Opin. Genet. Dev.* **4**: 25–30.

Schlessinger, J. (2000) Cell signaling by receptor tyrosine kinases. *Cell* **103**: 211–225.

Schlessinger, J. & Ullrich, A. (1992) Growth factor signaling by receptor tyrosine kinases. *Neuron* **9**: 383–391.

Schoeberl, B., Eichler-Jonsson, C., Gilles, E.D. & Muller, G. (2002) Computational modeling of the dynamics of the MAP kinase cascade activated by surface and internalized EGF receptors. *Nat. Biotechnol.* **20**: 370–375.

Scott, A.C. (1975) The electrophysics of a nerve fiber. *Rev. Mod. Phys.* **47**: 487–533.

Seet, B.T., Dikic, I., Zhou, M.M. & Pawson, T. (2006) Reading protein modifications with interaction domains. *Nat. Rev. Mol. Cell Biol.* **7**: 473–483.

Sewing, A., Wiseman, B., Lloyd, A.C. & Land, H. (1997) High-intensity Raf signal causes cell cycle arrest mediated by p21Cip1. *Mol. Cell Biol.* **17**: 5588–5597.

Sha, W., Moore, J., Chen, K., Lassaletta, A.D., Yi, C.S., Tyson, J.J. & Sible, J.C. (2003) Hysteresis drives cell-cycle transitions in *Xenopus laevis* egg extracts. *Proc. Natl Acad. Sci. USA* **100**: 975–980.

Shvartsman, S.Y., Muratov, C.B. & Lauffenburger, D.A. (2002) Modeling and computational analysis of EGF receptor-mediated cell communication in *Drosophila* oogenesis. *Development* **129**: 2577–2589.

Shymko, R.M., De Meyts, P. & Thomas, R. (1997) Logical analysis of timing-dependent receptor signalling specificity: application to the insulin receptor metabolic and mitogenic signalling pathways. *Biochem. J.* **326**: 463–469.

Taniguchi, C.M., Emanuelli, B. & Kahn, C.R. (2006) Critical nodes in signalling pathways: insights into insulin action. *Nat. Rev. Mol. Cell Biol.* **7**: 85–96.

Tischer, C. & Bastiaens, P.I. (2003) Lateral phosphorylation propagation: An aspect of feedback signalling? *Nat. Rev. Mol. Cell Biol.* **4**: 971–974.

Tyson, J.J., Chen, K. & Novak, B. (2001) Network dynamics and cell physiology. *Nat. Rev. Mol. Cell Biol.* **2**: 908–916.

Tyson, J.J., Chen, K.C. & Novak, B. (2003) Sniffers, buzzers, toggles and blinkers: dynamics of regulatory and signaling pathways in the cell. *Curr. Opin. Cell Biol.* **15**: 221–231.

Umbarger, H.E. (1956) Evidence for a negative-feedback mechanism in the biosynthesis of isoleucine. *Science* **123**: 848.

Van Obberghen, E., Baron, V., Delahaye, L. *et al.* (2001) Surfing the insulin signaling web. *Eur. J. Clin. Invest.* **31**: 966–977.

Wang, X., Hao, N., Dohlman, H. & Elston, T.C. (2006) Bistability, stochasticity and oscillations in the mitogen activated protein kinase cascade. *Biophys. J.* **90**: 1961-1978.

Werner, S.L., Barken, D. & Hoffmann, A. (2005) Stimulus specificity of gene expression programs determined by temporal control of IKK activity. *Science* **309**: 1857–1861.

Whitehead, J.P., Clark, S.F., Urso, B. & James, D.E. (2000) Signalling through the insulin receptor. *Curr. Opin. Cell Biol.* **12**: 222–228.

Whitehurst, A., Cobb, M.H. & White, M.A. (2004) Stimulus-coupled spatial restriction of extracellular signal-regulated kinase 1/2 activity contributes to the specificity of signal-response pathways. *Mol. Cell Biol.* **24**: 10145–10150.

Wolf, D.M. & Arkin, A.P. (2003) Motifs, modules and games in bacteria. *Curr. Opin. Microbiol.* **6**: 125–134.

Xiong, W. & Ferrell, J.E., Jr. (2003) A positive-feedback-based bistable 'memory module' that governs a cell fate decision. *Nature* **426**: 460–465.

Yarden, Y. & Sliwkowski, M.X. (2001) Untangling the ErbB signalling network. *Nat. Rev. Mol. Cell Biol.* **2**: 127–137.

Intracellular signalling during bacterial chemotaxis

Marcus J. Tindall, Philip K. Maini, Judy P. Armitage,
Colin Singleton and Amy Mason

1 Bacterial chemotaxis

Chemotactic bacteria are some of the best studied systems in biology. Their behaviour on the single cell and population scale has been investigated for the past 30 plus years both by experimentalists and theoreticians. Indeed work undertaken in this area is a paradigm of the success of mathematical modelling in helping to understand biological systems, even before the advent of 'systems biology'. However, although we know much of the biochemistry and basic characteristics of the systems, there remain a number of unanswered questions, both on the individual and collective population scale, about chemotactic bacteria.

Bacteria such as *E. coli* and *R. sphaeroides* respond to changes in extracellular attractant levels by changing the pattern of rotation of their flagella anchored across the cell membrane (Eisenbach *et al.*, 2004). In the case of *E. coli* changes in attractant concentration are detected by membrane-spanning methyl-accepting chemotaxis proteins (MCPs) at the poles of the rod-like shaped bacteria. Through a number of intracellular phosphotransfer biochemical reactions, these changes are communicated to the FliM protein motors that switch the direction of flagellar rotation. In the case of *E. coli*, the default setting of clockwise rotations in the absence of an attractant gradient, interspersed with periodic switching to clockwise rotation, leads to a series of run and tumbles. During such movement the individual four to six flagella rotate together to form a bundle which leads to short bursts of directed motion, followed by tumbling as a result of one or more flagella switching. Tumbling re-orientates the bacterium on a different directional heading each time. Considered over the order of minutes this movement leads to three-dimensional random-walk-like behaviour. When a change in the external concentration level of an attractant is detected by the MCPs, the bacterial flagella rotate for longer in a counter-clockwise direction leading to extended periods of runs up the attractant gradient resulting in chemotaxis.

If so much is known about bacterial chemotactic species, in particular *E. coli*, why do we continue to study them? The answer lies in the intriguing behaviour and characteristics these systems exhibit. For instance, *E. coli* are able to sense and respond to only small changes in attractant concentration (experimentally observed as small as the order of a few molecules; Segall *et al.*, 1986). This ability to magnify the effect of attractant binding in order to initiate the biochemical cascade between the receptors and the motors has been observed to be as high as 35 times the initial binding intensity (Sourjik and Berg, 2002b). Such a magnification is commonly referred to as gain and the mechanism for obtaining it has been of great interest to researchers for a number of years. Bacteria are also able to detect changes in concentration levels across five orders of magnitude, responding robustly in each case. Given the availability of data, both on the individual and population scale, mathematical models can be utilized to assist in explaining such observed characteristics.

Mathematical modelling has been particularly influential in the field of bacterial chemotaxis. A wide range of mathematical approaches have been used to help understand particular aspects of bacterial chemotactic systems, both on the individual and population scale (Tindall *et al.*, in press). Modelling work has often been undertaken in conjunction with experiment as we detail in Section 3. Mathematical models have been developed to explain experimental observations, verified against such observations and used to generate new hypotheses which are then experimentally tested. In other cases purely hypothetically based models have been formulated to explain biologically observed phenomena, for instance in the case of receptor–receptor interactions (Bray *et al.*, 1998).

In the work to be presented and discussed here we will focus specifically on the role of intracellular signalling within *E. coli*. A mathematical model will be developed to predict how the concentration of certain intracellular proteins affects the overall receptor-to-motor response of a bacterium. Importantly our modelling technique differs from previous models in that it seeks to simultaneously account for the effects that both spatial localization of the proteins and their predefined interactions have on the overall behaviour of the network. Generally, work in this area has focused primarily on modelling only the temporal dynamics, ignoring spatial aspects of the problem (Tindall *et al.*, in press). Our modelling technique will be compared with others to demonstrate both its usefulness and shortcomings.

Before presenting our specific modelling example we detail the known biological facts of the intracellular signalling pathway within *E. coli* in Section 2. Section 3 provides a brief overview of mathematical modelling in understanding various aspects of bacterial chemotaxis systems. In Section 4 we present a spatio-temporal mathematical model of the phosphotransfer pathway within *E. coli* specifically focusing on the proteins responsible for signalling between the membrane receptors and flagellar motors. Results from the model are presented and the effects of the outcomes discussed. The results of our work and the application of mathematical modelling in this and other fields of systems biology are discussed in Section 5.

2 Intracellular signalling within bacterial chemotaxis

Generally bacteria are too small in length (around 1–3 μm) to be able to detect spatial changes in their external attractant concentration. Instead they rely on an internal temporal system of biochemical signalling to communicate changes in the external environment via their MCPs to the flagella motors (Wadhams and Armitage, 2004). This internal intracellular network can range in complexity amongst various species of bacteria (Porter *et al.*, 2006; Rao *et al.*, 2004). In the case of *E. coli*, the internal phosphotransfer reaction is mediated by the activation of proteins associated with the

MCPs, primarily chemotaxis protein A (CheA) and CheW. CheW is thought to bind CheA to the cytoplasmic domain of the MCP receptor. In the absence of an attractant gradient, CheA autophosphorylates. The resulting phosphoryl groups are then transferred from the phosphorylated CheA protein molecules (CheA$_P$) to two proteins: CheB and CheY. CheY is an abundant protein within the bacterial cytoplasm. Diffusing from the receptors through the cytoplasm, phosphorylated CheY (CheY$_P$) is then free to interact with FliM protein motors which regulate the flagella switching. This interaction leads to the clockwise rotation of the flagella causing random walk behaviour. The protein CheZ actively dephosphorylates CheY$_P$ (CheY$_P$ also dephosphorylates naturally on a slower timescale).

The phosphoryl transfer from CheA$_P$ to CheB activates the esterase activity of CheB which demethylates the MCPs. This action of demethylation dynamically counteracts constant methylation of the receptors by the transferase protein CheR. Receptor methylation plays an important role in bacterial chemotaxis – it allows the system time to adapt to changes in different concentrations of the extracellular nutrient concentration. Whilst the initial change in motor bias is rapid (of the order of 100s of milliseconds), adaptation of the bacterium is relatively slow (of the order of up to 60 seconds). This period of delayed adaptation allows the bacterium time to undertake periods of extended runs and also attenuate its ability to detect new changes in the attractant concentration. It is through a careful balance of methylation of the receptors, which reduces sensitivity to bound attractant molecules, and demethylation, which increases sensitivity, that adaptation is achieved.

On detecting an increase in attractant concentration, through binding of attractant molecules to the receptors, the autophosphorylation of CheA is inhibited, thus leading to a drop in CheY$_P$ levels and subsequent counter-clockwise rotation of the flagella motors. Likewise CheB$_P$ levels fall and methylation of the receptors follows as a result of the constant activity of CheR. The increased methylation of the receptors finally leads to reactivation of the autophosphorylation of CheA, thereby resetting the system to its pre-stimulus levels. The details of the phosphotransfer process are summarized in *Figure 1* and each reaction is listed in *Table 1*.

3 Mathematical modelling and bacterial chemotaxis

Mathematical modelling is a useful and powerful tool for understanding and elucidating aspects of physical and biological systems. In order to build useful and informative models, certain aspects of the system must already be understood. A model can then be formulated and the respective model outcomes used to predict the behaviour of the system. Model outcomes are initially tested against known experimental results in order to verify the model. Certain attributes of the model (describing physical characteristics of the system) can then be altered to understand how they affect the outcomes, thus making predictions about the system's behaviour. Models can therefore be used to test experimental hypotheses and also direct future experimental work.

We note here that applied mathematical modelling does not constitute statistics or related fields such as bioinformatics, but instead focuses on understanding the 'system' by explaining its observed characteristics. Many aspects of such modelling are inherent in what is now commonly termed systems biology.

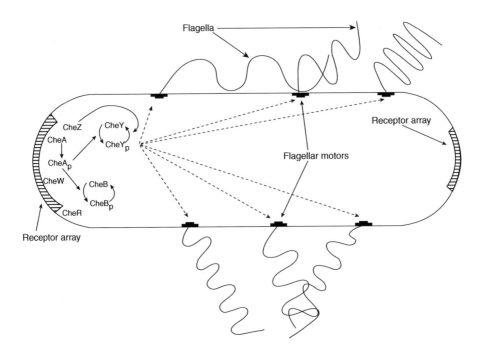

Figure 1. *A schematic representation (not to scale) of a typical* E. coli *bacterium (approximately 3 μm long) showing the location of the membrane receptors and flagella. CheW binds CheA to the cytoplasmic receptor domains. A decrease of attractant causes CheA to autophosphorylate, the phosphorylated CheA, CheA$_P$, passes the phosphoryl groups to both CheY and CheB. The subsequently phosphorylated CheY diffuses (indicated schematically by the dotted lines) to the flagellar motors where it causes them to rotate in a clockwise direction. CheY$_p$ can be dephosphorylated by CheZ. The phosphotransfer to CheB ensures that it acts to negate the methylating action of CheR, thus reducing the receptor methylation state. With the addition of attractant, the rate of autophosphorylation of CheA is reduced and thus CheY$_p$ levels fall. The motors respond to the drop in CheY$_p$ by rotating counter-clockwise causing the flagella to bundle together leading to runs. The dephosphorylation of CheB$_p$ allows CheR to further methylate the receptors thus inducing further CheA$_p$ activity, thereby returning the system to its pre-stimulus configuration.*

Many elements of the bacterial chemotaxis system, both on the individual and population scale, have been the subject of mathematical models for more than 40 years. The original application of mathematical modelling was used to understand the experimental assay work of Adler (Adler, 1966), who observed the migration of *E. coli* up gradients of attractant within a capillary tube (Keller and Segel, 1971). The modelling work focused on various different forms of the bacterial diffusion and chemotactic coefficients, thus seeking, in a somewhat coarse-grain fashion, to understand how microscale individual behaviour affected the macroscale population behaviour (Tindall *et al.*, in press).

Modelling of individual bacteria was motivated around the same time as the work of Adler by the work of Berg and colleagues (Block *et al.*, 1982, 1983), who observed

Table 1. *The autophosphorylation, phosphotransfer and dephosphorylation reactions within* E. coli.

Process	Reaction	Details
Autophosphorylation	$CheA \xrightarrow{\kappa_1} CheA_p$	
Phosphotransfer	$CheA_p + CheY \xrightarrow{k_2} CheA + CheY_p$	$CheA_p$ to CheY
	$CheA_p + CheB \xrightarrow{k_3} CheA + CheB_p$	$CheA_p$ to CheB
Dephosphorylation	$CheY_p + CheZ \xrightarrow{k_4} CheY + CheZ$	Dephosphorylation by CheZ
	$CheB_p \xrightarrow{k_5} CheB$	Natural dephosphorylation
	$CheY_p \xrightarrow{k_6} CheY$	Natural dephosphorylation

the attractant response behaviour of individual bacteria. Their subsequent findings on the interesting excitation and adaptation response of bacteria, and later work by Segall *et al.* (1986) on the highly sensitive response of the system to only small changes in the attractant concentration, has seen mathematical models used to understand particular elements of the bacterial system. Adaptation, sensitivity, gain or flagella dynamics, through to looking at the particular physical aspects of the bacterium, for example receptor dynamics, the phosphotransfer pathway, protein–motor interactions, and so on, have all been fertile areas. The growth in modelling sophistication has occurred in parallel with the ever-increasing understanding of the underlying biology. For instance, in the case of adaptation, early efforts focused on describing a number of basic receptor state models which relied on receptors being either attractant-bound or unbound, in either case moving between an inactive or active state (Goldbeter and Koshland, 1982; Segel and Goldbeter, 1986). Such inactive or active states were precursors to the discovery of the role of methylation in changing receptor activity, and later the proteins responsible for affecting this (CheR and $CheB_p$). With increasing understanding of the phosphotransfer pathway in the late 1980s and throughout the 1990s, the phosphotransfer pathway and role of CheR and $CheB_p$ in affecting methylation have been introduced to more recent models on adaptation and sensitivity (Barkai and Leibler, 1997; Spiro *et al.*, 1997).

The role of modelling has been particularly helpful in elucidating possible mechanisms responsible for sensitivity. With the advent of modelling of the phosphotransfer pathway within *E. coli* it was evident that a model of the pathway alone could not provide the necessary sensitivity and gain observed experimentally (Bray *et al.*, 1993; Spiro *et al.*, 1997). In using a discrete model in which receptors interact with one another to magnify the initial response of only one activated receptor through attractant binding, Bray and colleagues (Bray *et al.*, 1998) were able to show that this was a plausible mechanism for explaining both sensitivity and gain. Subsequent research, both experimental and theoretical, has focused on establishing the exact biochemical and physical forms of these interactions.

More recently, models of transport events within bacteria have focused on the role that the spatial environment within bacteria plays in affecting the phosphotransfer interaction between the receptors and flagellar motors (Lipkow et al., 2005; Lipkow, 2006). Here the effect of CheZ localization on the distribution of CheY$_p$ throughout the cytoplasm of the cell has been considered. Lipkow et al. (2005) found that when CheZ was allowed to diffuse throughout the cell, the CheY$_p$ concentration decreased exponentially from the receptor poles to the motors. However, when CheZ was localized to the membrane receptor regions of a cell, the concentration of CheY$_p$ was approximately constant throughout the cell, a result which agreed with experimental findings (Cantwell et al., 2003; Liberman et al., 2004).

With the growth of computing power and the sophistication of models both on the individual and population scale, the modelling of bacterial chemotaxis systems is heading towards more sophisticated multi-scale modelling techniques – where models on the individual scale are computed for a given population of cells (Bray et al., 2007, Emonet et al., 2005; Erban and Othmer, 2004; Kreft et al., 1998). The motivation for such work is to understand how individual cell behaviour affects the overall collective behaviour of the population. Such work has clear benefits, for example, in helping to understand biofilm behaviour, a consistent problem in medicine and the industrial world.

The more sophisticated a model on the individual cell level is, the more time it takes to integrate the respective input and provide output, thus when this is multiplied by the number of cells typical of a population the computational power required to provide answers in a feasible period of time will be very large. We are therefore faced with a 'double-edged' sword in that models must provide the relevant characteristics observed on the individual scale, but not become so sophisticated that understanding behaviour on the population scale becomes infeasible – so called 'model reduction'. Such work continues to provide challenges for mathematical modellers and experimentalists in seeking to understand bacterial chemotactic systems.

4 Developing a model of intracellular signalling

The focus of our work here is to understand how the spatio-temporal concentration of CheY$_p$ within a cell is dynamically affected by the removal of attractant from the bacterial receptors. What is the most appropriate mathematical modelling method for tackling this problem? It is worth noting that given the richness of mathematical theory and modelling approaches there are a number of choices. Should we consider modelling each individual protein molecule and its kinetics? How are we to include both the spatial and reaction processes simultaneously? Are stochastic effects important? Will our model be computationally efficient, that is, can it be solved in a reasonable period of time? How do we verify our model?

In order to develop an appropriate mathematical model we note that: (i) our theory needs to describe both the spatial localization of the respective proteins in time and a defined spatial region; and (ii) reaction rates and diffusion coefficients of the respective proteins are known from in vitro data (see Table 2). Given the reactions are known and the copy number (concentration) of each protein within the cytoplasm is high (see http://www.pdn.cam.ac.uk/groups/comp-cell/Rates.html for details), we can adopt an averaging approach in modelling individual protein molecules (both in space and time).

Table 2. *Dimensional and non-dimensional parameter values.*

Rate	Description	Value	Reference
k_1	Autophosphorylation of CheA	34 s^{-1}	Francis *et al.* (2002)
			Shrout *et al.* (2003)
k_2	Phosphotransfer from CheA$_P$ to CheY	1×10^8 (Ms)$^{-1}$	Stewart *et al.* (2000)
k_4	CheY$_P$ dephosphorylation by CheZ	1.6×10^6 (Ms)$^{-1}$	Li and Hazelbauer (2004)
			Sourjik and Berg (2002a)
k_6	CheY$_P$ natural dephosphorylation	8.5×10^{-2} s^{-1}	Smith *et al.* (2003)
			Stewart and van Bruggen (2004)
D_{Y_P}	CheY$_P$ diffusion coefficient	10 μm^2 s^{-1}	Elowitz *et al.* (1999)
			Segall *et al.* (1985)
A_T	Total CheA concentration in an *E. coli* cell	7.9 μm	Bray website data[1]
Y_T	Total CheY concentration in an *E. coli* cell	9.7 μm	Bray website data
Z	Total CheZ concentration in an *E. coli* cell	3.8 μm	Bray website data
$\bar{k}_2$	Non-dimensional phosphotransfer from CheAp to CheY.	28.53	–
$\bar{k}_4$	Non-dimensional CheYp dephosphorylation by CheZ.	0.179	–
$\bar{k}_6$	Non-dimensional CheYp natural dephosphorylation	2.5×10^{-3}	–
α	Ratio of total CheA to CheY concentration	0.814	–
$\bar{D}_Y$	Non-dimensional diffusion coefficient	1	–

[1] www.pdn.cam.ac.uk/groups/comp-cell/Rates.html

These points lead us to adopt the continuum mathematical theory of reaction-diffusion equations (Murray, 1993). Our system of governing partial differential equations (PDEs) will be of the form:

$$\frac{\partial v}{\partial t} = D_u \nabla^2 u + f(u,v), \tag{1}$$

$$\frac{\partial v}{\partial t} = D_v \nabla^2 u + g(u,v). \tag{2}$$

Here $u = u(\mathbf{x},t)$ and $v = v(\mathbf{x},t)$ are the concentrations of two proteins at spatial points $\mathbf{x} = (x,y,z)$ and time t, where the functions $f(u,v)$ and $g(u,v)$ describe the reactions between each protein. We have assumed the diffusion coefficients D_u and D_v are constant and isotropic and in the case of three-dimensional geometry

$$\nabla^2 = \frac{\partial^2}{\partial x^2} + \frac{\partial^2}{\partial y^2} + \frac{\partial^2}{\partial z^2}.$$

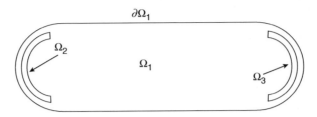

Figure 2. *A schematic representation of our two-dimensional model of an E. coli cell. Ω_2 and Ω_3 represent the regions near the receptor clusters in which CheA and CheA$_P$ are generally found. The outer boundary of the cell is denoted by $\partial\Omega_1$ and the remaining cytoplasmic region by Ω_1.*

This term (along with the diffusion coefficient) represents diffusion of the respective protein within the defined region of interest.

Given the symmetry of a cell along its length we only need to consider two-dimensional slices of the cytoplasmic region. Such a model can be solved computationally in a realistic time and in some cases, analytical approximations of the way in which the protein concentrations vary as a function of parameters within the model can be found.

Our system of equations is solved on the model cell shown in *Figure 2* based upon the following assumptions. We consider only phosphotransfer between CheA$_P$ and CheY$_P$ and the effect of CheZ. We neglect phosphotransfer to CheB given that it is concerned with local methylation of the receptors and not receptor-to-motor signalling. The role of CheB$_P$ will be included in future work as discussed in Section 5. We further assume that CheA, CheA$_P$ and CheZ are restricted to the two regions Ω_2 and Ω_3. CheA and CheA$_P$ are immobile in each region and the concentration of CheZ remains constant for all time. These regions are designed to represent the localization of CheA and CheA$_P$ to receptor clusters within the cell membrane. CheY and CheY$_P$ are free to diffuse throughout the full cytoplasmic region $(\Omega_1 \cup \Omega_2 \cup \Omega_3)$.

Invoking the law of mass-action (Murray, 1993) we can write down a set of reaction-diffusion equations governing each of the relevant reactions detailed in *Table 1* (neglecting CheB).

In the regions Ω_2 and Ω_3:

$$\frac{\partial A}{\partial t} = -k_1 A + k_2 A_p Y, \tag{3}$$

$$\frac{\partial A_p}{\partial t} = k_1 A - k_2 A_p Y, \tag{4}$$

$$\frac{\partial Y}{\partial t} = D_Y \nabla^2 Y - k_2 A_p Y + k_4 Y_p Z + k_6 Y_p, \tag{5}$$

$$\frac{\partial Y_p}{\partial t} = D_{Y_p} \nabla^2 Y_p + k_2 A_p Y - k_4 Y_p Z - k_6 Y_p, \tag{6}$$

and in Ω_1

$$\frac{\partial Y}{\partial t} = D_Y \nabla^2 Y + k_6 Y_P, \tag{7}$$

$$\frac{\partial Y_P}{\partial t} = D_{Y_P} \nabla^2 Y_P - k_6 Y_P. \tag{8}$$

Here $A = [\text{CheA}]$, $A_P = [\text{CheA}_P]$, $Y = [\text{CheY}]$, $Y_P = [\text{CheY}_P]$ and $Z = [\text{CheZ}]$ and the respective reaction rates and diffusion coefficients are detailed in *Table 2*. The initial distribution of CheA (see below) fills both Ω_2 and Ω_3, and neither CheA nor CheA$_P$ are assumed to diffuse within the respective regions (one is simply converted to the other and vice-versa) or outside them. CheZ is also assumed to remain constant within this region and thus we do not require an equation describing any changes in its concentration, either spatially or dynamically. The diffusion coefficients for CheY and CheY$_P$ are assumed to be the same in each of the three regions. We note here that given we are only considering a two-dimensional region:

$$\nabla^2 = \frac{\partial^2}{\partial x^2} + \frac{\partial^2}{\partial y^2}.$$

In order to complete or close our system of equations we need to define a set of initial and boundary conditions. Initial conditions define the initial protein concentration at time $t = 0$ in each region and boundary conditions detail the concentration on the boundary of the region (denoted as $\partial\Omega_1$ in *Figure 2*) for all time.

We assume initially that only ChcA is present in each of the receptor regions Ω_2 and Ω_3 and CheZ is constant throughout these regions. All other protein concentrations are zero. These conditions can be defined mathematically as follows.

In ω_1 we have

$$Y(\mathbf{x},0) = Y_0 \qquad \text{and} \qquad Y_P(\mathbf{x},0) = 0 \tag{9}$$

and in Ω_2 and Ω_3

$$A(\mathbf{x},0) = A_0, \ A_P(\mathbf{x},0) = 0, \ Y(\mathbf{x},0) = Y_0 \text{ and } Y_P(\mathbf{x},0) = 0. \tag{10}$$

Here A_0 is the initial concentration of CheA, Y_0 the initial concentration of CheY. For computational reasons an appropriately smooth function is chosen to represent A_0.

We assume no flux boundary conditions on $\partial\Omega_1$ which equates to neither CheY nor CheY$_P$ being able to diffuse outside the cell wall boundary

$$\hat{\mathbf{n}} \cdot \nabla Y(\mathbf{x},t) = 0 \text{ and } \hat{\mathbf{n}} \cdot \nabla Y_P(\mathbf{x},t) = 0. \tag{11}$$

Here $\hat{\mathbf{n}}$ is a unit vector normal to the surface of the cell. The flux of CheY and CheY$_P$ is taken to be continuous between each of the three regions Ω_1, Ω_2 and Ω_3.

4.1 Non-dimensionalization

Before attempting to solve the model, the dimensions from the governing equations are removed. Why do we do this? Non-dimensionalization refers to removing the dimensions from our equations. This allows consideration to be given to the effect that certain terms in our equations have in comparison to others. The relative magnitude of each can then be more clearly judged. Furthermore it reduces the number of parameters in our model.

Our governing equations contain both a lengthscale and a timescale and the governing equations are rescaled with respect to these. Here we non-dimensionalize with respect to the timescale of autophosphorylation of CheA (k_1) and the length scale associated with the diffusion distance of CheY$_P$ through the cell

$$t = \frac{\tau}{k_1} \quad \text{and} \quad \mathbf{x} = \sqrt{\frac{D_{Y_P}}{k_1}}\hat{\mathbf{x}}. \tag{12}$$

Here τ and $\hat{\mathbf{x}}$ are the non-dimensional timescales and length scales, respectively. The concentration of CheA, CheA$_P$, CheY and CheY$_P$ are rescaled with respect to the total phosphorylated and unphosphorylated CheA and CheY concentrations within the cell, A_T and Y_T, respectively

$$A = A_T\hat{A}, \qquad A_p = A_T\hat{A}_p, \qquad Y = Y_T\hat{Y} \quad \text{and} \quad Y_p = Y_T\hat{Y}_p. \tag{13}$$

If we substitute these rescalings into Equations (3)–(8) then in region Ω_1

$$\frac{\partial \hat{Y}}{\partial \tau} = \bar{D}_Y \nabla^2 \hat{Y} + \bar{k}_6 \hat{Y}_p, \tag{14}$$

$$\frac{\partial \hat{Y}_P}{\partial \tau} = \nabla^2 \hat{Y}_P - \bar{k}_6 \hat{Y}_p, \tag{15}$$

and in regions Ω_2 and Ω_3

$$\frac{\partial \hat{A}}{\partial \tau} = -\hat{A} + \bar{k}_2 \hat{A}_p \hat{Y}, \tag{16}$$

$$\frac{\partial \hat{A}_p}{\partial \tau} = \hat{A} - \bar{k}_2 \hat{A}_p \hat{Y}, \tag{17}$$

$$\frac{\partial \hat{Y}}{\partial \tau} = \bar{D}_Y \nabla^2 \hat{Y} - \alpha \bar{k}_2 \hat{A}_p \hat{Y} + (\bar{k}_4 + \bar{k}_6)\hat{Y}_p, \tag{18}$$

$$\frac{\partial \hat{Y}_P}{\partial \tau} = \nabla^2 \hat{Y}_p + \alpha \bar{k}_2 \hat{A}_p \hat{Y} - (\bar{k}_4 + \bar{k}_6)\hat{Y}_p, \tag{19}$$

where:

$$\bar{k}_2 = \frac{k_2 Y_T}{k_1}, \quad \bar{k}_4 = \frac{k_4 Z}{k_1}, \quad \bar{k}_6 = \frac{k_6}{k_1}, \quad \alpha = \frac{A_T}{Y_T} \text{ and } \bar{D}_Y = \frac{D_Y}{D_{YP}}, \tag{20}$$

are defined as the non-dimensional parameters. The number of parameters has been reduced from nine (k_1, k_2, k_4, k_6, D_Y, D_{YP}, A_T, Y_T, Z) to five ($\bar{k}_2, \bar{k}_4, \bar{k}_6, \bar{D}_Y, \alpha$).
The initial and boundary conditions are also rescaled such that in Ω_1,

$$\hat{Y}(\hat{\mathbf{x}}, 0) = 1 \qquad \text{and} \qquad \hat{Y}_p(\hat{\mathbf{x}}, 0) = 0 \tag{21}$$

and in Ω_2 and Ω_3

$$\hat{A}(\hat{\mathbf{x}}, 0) = 1, \quad \hat{A}_p(\hat{\mathbf{x}}, 0) = 0, \quad \hat{Y}(\hat{\mathbf{x}}, 0) = 1 \text{ and } \hat{Y}_p(\hat{\mathbf{x}}, 0) = 0 \tag{22}$$

with the boundary conditions now given by:

$$\hat{\mathbf{n}} \cdot \nabla \hat{Y}(\hat{\mathbf{x}}, t) = 0 \qquad \text{and} \qquad \hat{\mathbf{n}} \cdot \nabla \hat{Y}_p(\hat{\mathbf{x}}, t) = 0. \tag{23}$$

Here we have assumed that initially all of the CheY and CheA is unphosphorylated such that $Y_0 = Y_T$ and $A_0 = A_T$.

4.2 Parameterizing the model

To date we have formulated a spatio-temporal mathematical model of intracellular signalling in a bacterial cell based upon certain assumptions. Each term in the equations represents a certain process, for instance the reaction or diffusion of certain proteins. The parameters associated with each of these terms tell us (given their magnitude) the relative importance of each process. In the case of our *E. coli* cell parameterizing the system is relatively straightforward, the respective dimensional and corresponding non-dimensional parameters are listed in *Table 2*.
   The non-dimensional parameters show that in terms of importance phosphotransfer from CheA$_P$ to CheY is the dominant reaction process ($\bar{k}_2 = 28.53$), autophosporylation of CheA (our non-dimensionalization means this process is of order one), followed by dephosphorylation of CheY$_P$ by CheZ ($\bar{k}_4 = 0.179$) and finally the natural dephosphorylation of CheY$_P$ ($\bar{k}_6 = 2.5 \times 10^{-3}$).

4.3 Model solutions and results

We solve Equations (14)–(19), with their respective boundary and initial conditions, using the computational package COMSOL (Stockholm, Sweden). COMSOL uses the theory of finite elements to solve PDEs and allows dynamical behaviour to be studied in detail. *Figures 3A–D* (see colour plate section) show the change in concentration of CheY$_P$ at various time steps of the simulation. We note these results show the diffusion of CheY$_P$ from the receptor poles, where it is created, towards the centre of the cell. The CheY$_P$ diffuses through the cell at a rate dependent upon the magnitude of its diffusion coefficient.

5 Summary and future work

A spatio-temporal model of the phosphotransfer pathway within *E. coli* has been produced using the mathematical theory of reaction-diffusion equations. Solutions to the model have allowed us to examine the way in which the concentration of CheY$_P$ dynamically changes within the cell. The model results show the creation of CheY$_P$ at the receptor clusters which then diffuses towards the centre of the cell.

The modelling approach here differs from that of Lipkow *et al.* (2005), who undertook stochastic simulations involving each model protein. Lipkow *et al.* (2005) also considered the interaction between CheY$_P$ and the motor protein FliM, which we have not included here. However, although our approach has differed we have been able to reproduce the same findings in respect of the location of CheZ (results not shown). The model developed here will be extended further in future research to incorporate the CheB pathway. We will also use this approach to understand the importance of spatial protein localization in the more complex signal transduction system of *Rhodobacter sphaeroides* (Porter and Armitage, 2002).

This work demonstrates that although there are different approaches to modelling a biological problem, each theoretical technique can tell us something about the system that another may not be able to and *vice versa*. This demonstrates the strengths of theoretical models in being able to provide various insights into the biological system being analysed.

Although modelling approaches have been used to understand certain parts of the bacterial chemotaxis system, there still remain a number of unanswered questions, both on the individual and population scale. Of growing interest is how individual cell behaviour affects the overall growth and development of a population. Such work requires the use of multi-scale modelling (Tindall *et al.*, in press) which allows individual behaviour to be incorporated into population models. For instance, recent work by Bray *et al.* (2007) has demonstrated the importance of adaptation on the accumulation of bacteria in response to attractant gradients.

The application of mathematical models to problems in the biological sciences can provide valuable insight into the system being studied. It is important that models are based upon known biological evidence and are developed in strong collaborations between theoreticians and experimentalists. This approach leads to models being used to test certain hypotheses, helps direct future experimental work and thereby provide insight to the biological problem at hand. In short these points are the real strengths of systems biology.

Acknowledgements

This work (MJT) was funded by a grant (BB/C513350/1) from the Biotechnology and Biological Sciences Research Council (BBSRC), UK. PKM was partly supported by a Royal Society Merit Award.

References

Adler, J. (1966) Chemotaxis in bacteria. *Science* **153:** 708–716.
Barkai, N. & Leibler, S. (1997) Robustness in simple biochemical networks. *Nature* **387:** 913–917.

Block, S., Segall, J. & Berg, H. (1982) Impulse response in bacterial chemotaxis. *Cell* **31:** 215–226.

Block, S., Segall, J. & Berg, H. (1983) Adaptation kinetics in bacterial chemotaxis. *J. Bacteriol.* **154:** 312–323.

Bray, D., Bourret, R. & Simon, M. (1993) Computer simulation of the phosphorylation cascade controlling bacterial chemotaxis. *Mol. Biol. Cell* **4:** 469–482.

Bray, D., Levin, M. & Morton-Firth, C. (1998) Receptor clustering as a cellular mechanism to control sensitivity. *Nature* **393(7):** 85–88.

Bray, D., Levin, M. & Lipkow, K. (2007) The chemotactic behavior of computer-based surrogate bacteria. *Curr. Biol.* **17:** 12–19.

Cantwell, B., Draheim, R., Weart, R., Nguyen, C., Stewart, R. & Manson, M. (2003) CheZ phosphatase localizes to chemoreceptor patches via CheA-short. *J. Bacteriol.* **185(7):** 2354–2361.

Eisenbach, M., Lengeler, J., Varon, M. *et al.* (2004) *Chemotaxis.* Imperial College Press, London.

Elowitz, M., Surette, M., Wolf, P., Stock, J. & Leibler, S. (1999) Protein mobility in the cytoplasm of *Escherichia coli. J. Bacteriol.* **181(1):** 197–203.

Emonet, T., Macal, C., North, M., Wickersham, C. & Cluzel, P. (2005) Agentcell: a digital single-cell assay for bacterial chemotaxis. *Bioinformatics* **21(11):** 2714–2721.

Erban, R. & Othmer, H. (2004) From individual to collective behaviour in bacterial chemotaxis. *SIAM J. Appl. Math.* **65:** 361–391.

Francis, N., Levit, M., Shaikh, T., Melanson, L. & Stock, J. (2002) Subunit organization in a soluble complex of Tar, CheW and CheA by electron microscopy. *J. Biol. Chem.* **277:** 36755–36759.

Goldbeter, A. & Koshland, D. (1982) Simple molecular model for sensing and adaptation based on receptor modification with application to bacterial chemotaxis. *J. Mol. Biol.* **161:** 395–416.

Keller, E. & Segel, L. (1971) Travelling bands of chemotactic bacteria: A theoretical analysis. *J. Theor. Biol.* **30(2):** 235–248.

Kreft, J., Booth, G. & Wimpenny, J. (1998) Bacsim, a simulator for individual-based modelling of bacterial colony growth. *Microbiology* **144:** 3275–3287.

Li, M. & Hazelbauer, G. (2004) Cellular stoichiometry of the components of the chemotaxis signalling complex. *J. Bacteriol.* **186:** 3687–3694.

Liberman, L., Berg, H. & Sourjik, V. (2004) Effect of chemoreceptor modification on assembly and activity of the receptor-kinase complex in *Escherichia coli. J. Bacteriol.* **186:** 6643–6646.

Lipkow, K. (2006) Changing cellular location of CheZ predicted by molecular simulations. *PLoS Comput. Biol.* **2(4):** 301–310.

Lipkow, K., Andrews, S. & Bray, D. (2005) Simulated diffusion of phosphorylated CheY through the cytoplasm of *Escherichia coli. J. Bacteriol.* **187(1):** 45–53.

Murray, J. (1993) *Mathematical Biology,* 2nd Edn. Springer Verlag, Berlin.

Porter, S. & Armitage, J. (2002) Phosphotransfer in *Rhodobacter sphaeroides* chemotaxis. *J. Mol. Biol.* **324:** 35–45.

Porter, S., Wadhams, G., Martin, A., Byles, E., Lancaster, D. & Armitage, J. (2006) The CheYs of *Rhodobacter sphaeroides. J. Biol. Chem.* **281(43):** 32694–32704.

Rao, C., Kirby, J. & Arkin, A. (2004) Design and diversity in bacterial chemotaxis: a comparative study in *Escherichia coli* and *Bacillus subtilis.* PLoS Biol. **2(2):** 239–252.

Segall, J., Ishihara, A. & Berg, H. (1985) Chemotactic signalling in filamentous cells of *Escherichia coli. J. Bacteriol.* **161(1):** 51–59.

Segall, J., Block, S. & Berg, H. (1986) Temporal comparisons in bacterial chemotaxis. *Proc. Natl Acad. Sci.* USA **83(23):** 8987–8991.

Segel, L. & Goldbeter, A. (1986) A mechanism for exact sensory adaptation based on receptor modification. *J. Theor. Biol.* **120:** 151–179.

Shrout, A., Montefusco, D. & Weis, R. (2003) Template-directed assembly of receptor signalling complexes. *Biochemistry-US* **42(46):** 13379–13385.

Smith, J., Latiolais, J., Guanga, G., Citineni, S., Silversmith, R. & Bourret, R. (2003) Investigation of the role of electrostatic charge in activation of the *Escherichia coli* response regulator CheY. *J. Bacteriol.* **185(21):** 6385–6391.

Sourjik, V. & Berg, H. (2002a) Binding of the *Escherichia coli* response regulator CheY to its target measured in vivo by fluorescence resonance energy transfer. *Proc. Natl Acad. Sci.* USA **99:** 12669–12674.

Sourjik, V. & Berg, H. (2002b) Receptor sensitivity in bacterial chemotaxis. *Proc. Natl Acad. Sci.* USA **99(1):** 123–127.

Spiro, P., Parkinson, J. & Othmer, H. (1997) A model of excitation and adaptation in bacterial chemotaxis. *Proc. Natl Acad. Sci.* USA **94:** 7263–7268.

Stewart, R. & van Bruggen, R. (2004) Rapid phosphotransfer to CheY from a CheA protein lacking the CheY-binding domain. *Biochemistry-US* **43(27):** 8766–8777.

Stewart, R., Jahreis, K. & Parkinson, J. (2000) Rapid phosphotransfer to CheY from a CheA protein lacking the CheY-binding domain. *Biochemistry-US* **39(43):** 13157–13165.

Tindall, M., Maini, P., Porter, S. & Armitage, J. (2007a) Overview of mathematical approaches used to model bacterial chemotaxis II: Bacterial populations. *Bull. Math. Biol.*, In press.

Tindall, M., Porter, S., Maini, P., Gaglia, G. & Armitage, J. (2007b) Overview of mathematical approaches used to model bacterial chemotaxis I: The single cell. *Bull. Math. Biol.*, In press.

Wadhams, G. & Armitage, J. (2004) Making sense of it all: Bacterial chemotaxis. *Nat. Rev. Mol. Cell Biol.* **5(12):** 1024–1037.

9

Modelling the mammalian heart

Richard Clayton and Martyn Nash

1 Introduction

The human heart is a muscular organ, about the size of a fist, and its function can be considered at different length scales. At the organ scale, regular and synchronized contractions of the heart wall propel blood around the vascular system, supplying oxygen and maintaining homeostasis within the tissues of the body. Each heart beat at the tissue scale is associated with a propagating wave of electrical depolarization and contraction, followed by electrical repolarization and relaxation. At the scale of a single cell, electrical depolarization results from the entry of Na^+ and Ca^{2+} into the cell through ion channels embedded in the cell membrane, and the ensuing increase in intracellular Ca^{2+} concentration triggers release of Ca^{2+} from intracellular stores, which in turn activates a sliding filament mechanism that produces tension. At the molecular scale, entry of Na^+ into the cell during depolarization is tightly controlled by the conformation of ion channel proteins embedded in the cell membrane, which alters in response to local electric field strength.

These scales are linked, because organ level function of the heart is an emergent product of events at the molecular scale. An example of these links that we will consider later is gene polymorphisms associated with altered function of the Na^+ ion channel. These molecular level changes result in organ level phenomena including cardiac arrhythmias. However, the causal chain operates in both directions, and events at the tissue and organ scale also influence the behaviour of cells and molecules. For example, mechanical stress within the heart wall can open stretch activated ion channels, and so a tissue level effect influences molecular level processes. This bi-directional coupling across scales poses a problem for explaining heart function; should the explanation start with small scale (molecule and cell) detail and work upwards to describe the tissue and organ scales, or should it start with tissue and organ scale function and infer the underlying cellular and molecular mechanisms that are important? This issue of whether to take a 'bottom-up' or 'top-down' approach is a general problem in systems biology. A bottom-up approach relies on detailed and complete knowledge of small scale function, whereas a top-down approach does not guarantee a unique explanation.

This important and fundamental issue is considered in depth elsewhere (Noble, 2006), and the aim of this chapter is to describe how this issue has been tackled when modelling the mammalian heart. We start with an overview of the structure and function of cardiac tissue, and then move on to describe the different components of a heart model, and the numerical and computational issues associated with solving them. Finally, we briefly describe two success stories where a modelling approach has been used to gain a significant insight into the properties and behaviour of cardiac tissue.

2 Physiological background

2.1 Structure and function of the heart

The mammalian heart is a pump with four chambers; two atria that collect venous blood, and two ventricles that pump blood to the arterial system. The mechanical activity of the heart is synchronized so that the atria contract first, acting to prime the ventricles, which then contract after a short delay. The regular, spontaneous activity, as well as the synchronization, is achieved by propagating waves of electrical activation, the action potential, which initiate contraction via the ubiquitous second messenger Ca^{2+}.

2.2 Cardiac cells

Cardiac cells are generally cylindrical in shape, about 30–100 μm long and 8–20 μm in diameter, depending on their location on the heart. Ion channel and exchanger proteins embedded within the impermeable membrane allow the passage of ions between the extracellular and intracellular space, and gap junction proteins permit the flow of ions between the intracellular spaces of neighbouring cells. The cardiac action potential results from the movement of Na^+, Ca^{2+}, and K^+ ions through these proteins, and is controlled by their voltage, time, and concentration-dependent properties.

Within the cell lie two structures that are vital for electromechanical function. The sarcoplasmic reticulum (SR) acts as an intracellular Ca^{2+} store. Pumps for Ca^{2+} uptake and localized sites (ryanodine receptors) for Ca^{2+} release are embedded in the SR membrane, and maintain a high concentration of SR Ca^{2+} relative to the cytosol. The second major structure is the sliding filament mechanism that generates tension within the cell.

2.3 The action potential

At rest, cardiac cells maintain a lowered cytosolic concentration of both Na^+ and Ca^{2+}, and an elevated cytosolic concentration of K^+ compared to the extracellular space. These differences in concentration produce a potential difference of around –90 mV across the cell membrane, with the intracellular space being polarized at a lower potential than the extracellular space. During an action potential, Na^+ and Ca^{2+} move into the cell resulting in depolarization of the membrane potential to between +10 and +50 mV. This change in membrane potential opens K^+ channels, and the movement of K^+ out of the cell results in repolarization of the cell membrane to its resting value of around 90 mV. During and after repolarization, pumps and exchangers in the cell membrane return the cytosolic and extracellular concentrations of Na^+, Ca^{2+}, and K^+ to their resting values.

Each heart beat therefore originates as a spontaneous action potential that arises in the sinoatrial node (SAN), which acts as the heart's natural pacemaker. In the human heart, the period of this spontaneous activity is about one second. It is important to note that the periodicity itself is an integrative phenomenon arising from interactions between membrane voltage and the different membrane currents present in the SAN, and not the result of an underlying oscillator (Noble, 2006). The action potential propagates first into the atria, then slowly through the atrioventricular node (AVN) with a delay of around 150 ms, quickly through the specialized conduction system of the ventricles, and finally from the ventricular endocardium to the ventricular epicardium.

The changes in potential produced by a propagating action potential are also manifest as potential changes on the body surface because the body acts as a volume conductor. These potentials are the electrocardiogram (ECG), which can be measured using electrodes placed on the body surface, and the ECG is the standard noninvasive clinical tool for assessing the electrical function of the heart. The biophysical concepts underlying the origin of the ECG are covered in detail in Plonsey and Barr (2000).

2.4 *Cardiac tissue and action potential propagation*

An action potential propagates along the cell membrane because of depolarizing local current flow ahead of the action potential upstroke. This depolarization opens Na^+ channels, and hence initiates a propagating action potential. Cardiac cells are coupled to each other by gap junctions, and so the action potential propagates from one cell to its neighbours. The rod-like cardiac cells are aligned in fibres, which are organized in sheets of four to six cells thickness (LeGrice *et al.*, 1995). This orthotropic structure of ventricular muscle gives rise to excitation propagation speeds that differ with direction, whether it be along the fibres (the 'fibre' axis), across fibres within sheets ('sheet' axis), or across the sheets ('sheet-normal' axis). The density of gap junctions is highest at the ends of cardiac cells, and consequently an action potential propagates about twice as fast along the fibre axis than across it.

2.5 *Ca^{2+}-mediated coupling of electrical and mechanical activity*

The entry of Ca^{2+} into the cell during the early phase of the action potential triggers release of Ca^{2+} from the SR stores, resulting in an increase of Ca^{2+} in the cytosol. The contractile machinery is activated by the binding of Ca^{2+} to the filament protein Troponin C, and so the release of Ca^{2+} leads to tension development within the cell. Following Ca^{2+} release from the SR, Ca^{2+} is removed from the cytosol by mechanisms including exchangers that remove Ca^{2+} from the cell and pumps that return Ca^{2+} to the SR store. A detailed review of the role of Ca^{2+} in cardiac cells can be found in Bers (2002).

2.6 *Tissue mechanics*

The fibrous and laminar structure of cardiac tissue described above also gives rise to orthotropic passive mechanical properties, for which the stiffness is greatest along the fibre axis due to the tightly bound endomysial collagen that surrounds muscle cells. On the other hand, the sheets of myocytes can slide over each other with relative ease due their relatively loose coupling by perimysial collagen. The systolic contraction phase of the heart cycle is driven by active tension that is generated primarily along the axes of the fibres. During systole, shearing of the myocardial sheets has been shown to account for the wall thickening (LeGrice *et al.*, 1995).

2.7 Mechano-electrical feedback

As well as feed-forward from the action potential to contraction of the cardiac cell, there is also feed-back from mechanical to electrical activity. One of the most dramatic examples of this phenomenon is *commotio cordis*, where a mechanical blow to the chest can initiate an electrical cardiac arrhythmia. Stretch-activated ion channels in the cell membrane underlie mechano-electrical feedback, and are reviewed elsewhere (Kohl and Sachs, 2001).

2.8 Role of modelling

Our understanding of the structure and function of the heart described above has been obtained through dedicated experimental work over many years, which has resulted in detailed knowledge of, for example, the structure and function of individual ion channels. Given the enormous progress in experimental techniques that have been seen in recent years it is legitimate to ask why modelling the heart is important and worthwhile. The answer to this question lies partly in the fundamentally integrative nature of biological systems. In order to understand the propagation of an action potential in cardiac tissue it is important to take into account the electrical behaviour of the cells that compose the tissue, as well as organ level events such as mechanical strain in the heart wall. A mathematical model (encoded as a computational model) is an excellent way to describe, explain, and understand this type of system.

Later on in this chapter, we will describe the use of models for investigating the mechanisms that initiate and sustain lethal cardiac arrhythmias. These phenomena are extremely difficult to study in the human heart for entirely appropriate ethical reasons. While they can be investigated using animal tissue, observations of electrical activation patterns using contact electrodes or voltage sensitive fluorescent dyes are mainly limited to the tissue surface. In the thick walled ventricles, activation patterns are three-dimensional, and models have proved crucial in interpreting experimental data from the surface. A helpful analogy is study of the sun. It is impossible to make direct measurements within the sun's interior, because it is so hot. The sun and other stars are studied by observing their surface features and the radiation they emit, and interpreting these observations with models of stellar nuclear reactions.

Physical scientists and engineers routinely use models to examine complex systems and to predict their behaviour. In these areas it is an approach that requires no justification at all. A structure such as a bridge or a dam will be designed using a model of the mechanical stresses and strains generated in the load-bearing parts of the structure. In this way a design can be developed and finely tuned using the model, without having to build the full size structure. The approach that is taken here is not to describe the mechanical behaviour of the structure in terms of the bonds between individual molecules. The model is constructed to determine whether the building is safe, and so the mechanical properties of small regions are lumped together, and the behaviour of the entire structure is assessed by modelling the interactions between these finite elements. The size of the elements has an important influence on the fidelity of the results, and it will often be necessary to use smaller elements close to regions that are under greatest mechanical load. The main point here is that the complexity and level of detail in the model is adjusted so that the model is able to answer a specific question, such as whether a building will collapse during an earthquake of particular magnitude.

The heart is more complex than a mechanical structure. As we have seen, the electrical and mechanical properties of cardiac tissue are non-linear and anisotropic, and furthermore there is strong coupling between structure and function at the cell, tissue, and organ levels. A modelling approach does allow these multi-scale interactions to be unpicked, but in order to produce a useful and computationally tractable model the amount of detail is adjusted so that a specific research question can be addressed. In the following section, we describe the range of models available at the cell, tissue and organ level, before going on to describe how they can be assembled. A more detailed explanation of this overall approach can be found elsewhere (Crampin *et al.*, 2004; Noble, 2006), and the interested reader is also directed to detailed reviews of electrical (Rudy, 2006) and electromechanical (Kerchoffs *et al.*, 2006) models.

3 Cell models

3.1 *Electrical models*

Throughout the second half of the 20th century there was a strong interaction between experimentalists and modellers of cardiac cells. Each new experimental technique produced data, which increased the level of detail captured by models of excitable cells. At the same time, models were used as a tool to generate hypotheses about the possible existence of additional ion currents. This fascinating story has been described in detail elsewhere (Noble and Rudy, 2001), and what follows aims to capture the main developments.

The foundations of cardiac cell models were laid by Hodgkin and Huxley in their hugely influential study of the squid giant axon (Hodgkin and Huxley, 1952). Painstaking experimental work was combined with a mathematical model to explain both the shape and propagation of action potentials in the squid axon. The basis of their approach was an electrical model of the cell membrane where the current flow through each type of ion channel was modelled as a time-and voltage-dependent conductance in series with the Nernst potential for that ionic species. This approach was first adapted for cardiac Purkinje cells by Noble (1962), and the general mathematical form of these models is as follows:

$$C_m \frac{dV_m}{dt} = -\sum_S I_S \qquad (1)$$

$$I_S = \overline{G_S} \, x_S^1 \, x_S^2 \dots x_S^n \, (V_m - V_S) \qquad (2)$$

$$\frac{dx_S}{dt} = \frac{x_{S\infty}(V_m) - x_S}{\tau_S(V_m)}. \tag{3}$$

Here V_m is the transmembrane voltage, C_m the specific membrane capacitance (capacitance per unit area), I_S the current carried through a particular type of ion channel by ionic species S, $\overline{G_S}$ the maximal conductance of the ion channel, x_S gating variables that vary between 0 and 1 and describe the activation, inactivation and recovery of the ion channel, V_S the Nernst potential of species S, $x_{S\infty}$ the steady-state value of a single gating variable x_S (i.e. its value for $dx/dt = 0$), and τ_S a time constant describing the return of x to its steady-state value x_∞ following a voltage perturbation. The voltage dependence of both x_∞ and τ_x can be determined from experimental data (Hodgkin and Huxley, 1952; Rudy, 2006).

The Noble 1962 model was adapted for cardiac ventricular tissue by Beeler and Reuter (1977), and this model is illustrated in *Figure 1*. There are four types of current. I_{Na} is the dominant current during the rapid depolarization of the action potential upstroke, activating quickly and reaching a peak value of around -1 μA mm^{-2} before inactivating equally rapidly. The other inward (negative) current is I_S, which models the influx of Ca^{2+} ions. The maximal conductance of this channel, $\overline{G_S}$, depends on the intracellular Ca^{2+} concentration, which is modelled by a single ordinary differential equation. The other two currents I_{K1} and I_K act to repolarize the membrane, and are outward (positive) currents.

Experimental data suggested that the I_{K1} current in this model was actually composed of several components, and the model was revised and expanded by Luo and Rudy (1991). However, in this model as in other first generation ionic models, changes in intracellular Ca^{2+} concentration were modelled in a very simple way. Second generation ionic models include detailed models of the balance of each ionic species, and how the extracellular and intracellular concentration is affected by the ion channels, pumps and exchanger currents that carry it. Some second generation models also include an additional description of the subspace and cleft space close to the intracellular and extracellular surface of the membrane. The earliest second generation model was the DiFrancesco–Noble model of Purkinje cells (DiFrancesco and Noble, 1985), and other early second generation models of ventricular cells include the Luo–Rudy dynamic (LRd) model (Luo and Rudy, 1994) and the Noble 1998 model (Noble *et al.*, 1998) for guinea pig ventricular cells. There are now many second generation ventricular cell models including models for rat (Pandit *et al.*, 2001), rabbit (Puglisi and Bers, 2001), guinea pig (Luo and Rudy, 1994), canine (Fox *et al.*, 2002), and human (TenTusscher *et al.*, 2004).

This short section has concentrated on ventricular cell models. More comprehensive reviews covering other cell types including the sinoatrial node can be found elsewhere (Kleber and Rudy, 2004; Rudy, 2006).

3.2 *Mechanical models*

The electrophysiology of cardiac myocytes leads to contraction as a result of calcium ions being released from the sarcoplasmic reticulum (SR) during each cycle. This is

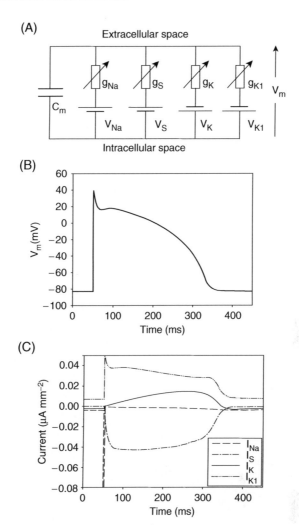

Figure 1. *Beeler-Reuter cardiac cell model, showing (A) equivalent circuit of the cell membrane, (B) example action potential, and (C) current flow during the action potential.*

followed by relaxation associated with the uptake of calcium back into the SR. Mechanisms of muscle contraction were identified 50 years ago with the formulation of the sliding filament theory of muscle mechanics (Huxley, 1957). The basic contractile unit within the cell is the sarcomere, which is packed with overlapping contractile proteins known as the thick (myosin) and thin (actin) filaments. Projections between the filaments known as cross-bridges generate tension and consume energy (ATP) as they rapidly cycle (attach and detach; Huxley and Simmons, 1971). This leads to shortening of the sarcomeres and ultimately myocyte contraction.

Detailed mathematical models based on the sliding filament theory have been developed over the decades (Guccione *et al.*, 1993; Smith, 1990; Zahalak, 2000). However, whilst these so-called 'distribution-moment' models reliably reproduce the detailed intracellular biophysics of contraction, their use in larger scale (tissue and organ) simulations is hampered because of their computational demand. Based on continuum mechanics assumptions, a more compact and computationally tractable 'fading-memory' model approach was developed in order to reproduce the key features of cellular contraction including calcium binding to actin filaments, availability of actin binding sites, and the kinetics of cross-bridge cycling (Hunter *et al.*, 1998). The parameters of this fading-memory model were recently refitted following a comprehensive review of cardiac muscle mechanics experiments (Niederer *et al.*, 2006). The variables that modulate active tension as inputs to this model are the sarcomere length (which is typically the result of solving an elasticity problem – see below), and the intracellular calcium concentration, which is typically a state variable in the electrophysiology models. These links provide a framework for excitation-contraction coupling.

4 Cardiac tissue models

4.1 *Electrical models*

A consideration of the passive and active behaviour of an infinite one-dimensional fibre with an excitable membrane leads to the cable equation that was used by Hodgkin and Huxley (1952) to simulate action potential propagation in the squid axon. This approach can be generalized to two and three dimensions, where cardiac tissue is considered as a two-phase medium composed of intracellular and extracellular spaces. The anisotropic electrical properties of cardiac tissue are described by conductivity tensors. The transmembrane potential is the difference between the intracellular and extracellular potentials:

$$V_m = \Phi_i - \Phi_e \tag{4}$$

and the overall system of equations is the following:

$$\nabla \cdot (\mathbf{G}_i + \mathbf{G}_e)\Phi_e = -\nabla \cdot (\mathbf{G}_i \nabla V_m) \tag{5}$$

$$\nabla \cdot (\mathbf{G}_i \nabla V_m) + \nabla \cdot (\mathbf{G}_i \nabla \Phi_e) = -S_v I_m \tag{6}$$

$$I_m = C_m \frac{dV_m}{dt} + I_{ion}. \tag{7}$$

Here ∇ is the gradient operator, $\nabla \cdot$ the divergence operator, the subscripts i and e denote intracellular and extracellular spaces, respectively, I_m is current flow through the membrane per unit area, $\mathbf{G}$ conductivity tensors, and S_v the surface to volume ratio for cells. I_m includes the current flow through ion channels, pumps and exchangers I_{ion}, which can be obtained from the cell models described above. This bidomain model includes an explicit description of intracellular and extracellular spaces, and the differing anisotropy of these spaces is modelled by having different conductivity tensors for the intracellular and extracellular space. This feature enables phenomena such as current injection into tissue to be modelled in detail. However, the bidomain equations are time consuming to solve numerically, and if the conductivity tensors $\mathbf{G}_i$ and $\mathbf{G}_e$ are equal or proportional to each other (i.e. the anisotropy in each domain is the same), substitution of Equations (5) and (7) into (6) leads to the monodomain equation:

$$\nabla \cdot (\mathbf{G} \nabla V_m) = S_v \left[C_m \frac{\partial V_m}{\partial t} + I_{ion} \right]. \tag{8}$$

This equation is usually rearranged and written in the form of a reaction-diffusion partial differential equation:

$$\frac{\partial V_m}{\partial t} = \nabla \cdot (\mathbf{D} \nabla V_m) - \frac{I_{ion}}{C_m} \tag{9}$$

where the diffusion tensor $\mathbf{D}$ is $\mathbf{G}/S_v C_m$ and has units of distance^2 time^{-1}. This equation is less time consuming to solve than the bidomain equations, and provides a good compromise between biophysical detail and computational tractability for modelling propagation.

4.2 Mechanical models

Newton's laws of motion provide the fundamental continuum mechanics framework for modelling deformations of soft biological tissues (Malvern, 1969) and the analysis is typically performed using finite element methods (Oden, 1972). Neglecting external body accelerations and body forces, the conservation of linear momentum can be reduced to the governing equations for static equilibrium, which state that the divergence of mechanical stress is zero:

$$\frac{\partial \sigma_{ij}}{\partial x_i} = 0 \tag{10}$$

where σ_{ij} is the Cauchy stress tensor, and x_i are the geometric coordinates of the deformed configuration. Given that heart tissue is primarily made up of water, it is typically appropriate to assume that myocardium is incompressible in nature, particularly when dealing with quasi-static (steady-state) analyses. Incompressibility imposes additional kinematic constraints on mechanical function of the tissue. Critical to the mechanical analysis of cardiac tissues is the use of finite deformation theory, which results in non-linear systems of governing equations. Adding to this complexity are the non-linear mechanical properties of myocardium. The choice of a suitable constitutive relation to relate the mechanical stress to the tissue deformation is essential for reliable modelling of cardiac mechanics.

Often the key challenge involves the choice of the stress-strain relation in order to encapsulate the passive non-linear anisotropic mechanical response of myocardial tissue. A variety of hyperelastic myocardial constitutive relations have been formulated for ventricular muscle in terms of the Green strains, the strain invariants, entropy considerations, and so on (Costa *et al.*, 2001; Guccione *et al.*, 1991; Humphrey *et al.*, 1990a; Nash and Hunter, 2000). The difficulty with all of these relations is the process of material parameter identification. Some studies have used data from biaxial tension tests (Humphrey *et al.*, 1990b), however, the collagen disruption and tissue damage resulting from the thin sectioning of the tissues brings into question the suitability of the derived parameters for *in vivo* modelling. Simple shear experiments on 3-D blocks of myocardium (Dokos *et al.*, 2002) were used to compare the performance of a variety of orthotropic relations (Schmid *et al.*, 2007), and it was found that the orthotropic Fung-type exponential relation outperformed the others in terms of parameter determinability, goodness of fit, and parameter variability. This relation was also used successfully in a 3-D *in vivo* analysis of myocardial mechanical properties (Augenstein *et al.*, 2005), and provides a compact, efficient constitutive relation suitable for use in large-scale models of cardiac mechanics.

5 Whole-organ models

5.1 *Anatomical models*

Whole-organ anatomical models include the 3-D geometry, together with a description of the orientation of fibres and local cell type. It is possible to use simplified models to investigate generic features of propagation, where, for example, ventricular geometry is abstracted to a slab representing a piece of the ventricular wall.

Although a whole-organ model is a worthwhile goal, the current emphasis is on detailed models of different parts of the heart. Abstract models based on idealized atrial and ventricular anatomy have been developed, but the emphasis has been on anatomical models obtained from real specimens. Most activity has concentrated on detailed models of the ventricles, but anatomical models of other regions have been developed, notably the sinoatrial node where models of structure and function have been used to investigate how regional differences in coupling and electrophysiology contribute to the normal pacemaking activity of the heart (Dobrzynski *et al.*, 2005).

For ventricular anatomy, high-resolution dissection and histology has been used to determine both geometry and fibre orientation in the rabbit (Vetter and McCulloch, 1998), canine (Nielsen *et al.*, 1991), and pig (Stevens *et al.*, 2003) ventricle. In these models, the measurements were fitted to a finite element description where the finite elements describe the 3-D geometry, and the fibre orientation was expressed as a set of direction cosines that varied smoothly as a field within each element. With the development of high-strength magnets offering high-resolution magnetic resonance imaging (MRI), and techniques such as diffusion tensor MRI (Le Bihan *et al.*, 2001), it has become possible to obtain anatomically detailed models without dissection because these techniques provide information on fibre orientation as well as anatomy. *Figure 2* (see colour plate section) illustrates such an anatomical model obtained from diffusion tensor MRI by the Johns Hopkins group (Helm *et al.*, 2005). Automated segmentation from these data can be difficult because the greyscale does not distinguish between excitable and connective tissue and the greyscale may in addition be non-uniform between slices. One solution to this problem is to combine high-resolution MRI with histology data from the same specimen, enabling precise and guided segmentation. Recent studies show that this approach is capable of producing very finely detailed models of the rabbit ventricle (Burton *et al.*, 2006).

5.2 *Whole-organ electrical models*

The earliest whole ventricle electrical models were based on relatively low resolution geometries, and greatly simplified cell and tissue models (Wei, 1997). These models were used to investigate excitation and recovery patterns during normal beats, and these insights have informed the development of torso models that enable potentials on the ventricular myocardium to be estimated from potentials on the body surface (Burnes *et al.*, 2001; Rudy, 2000). A more widespread and recent application of whole-ventricle models has been to study the mechanisms of both fibrillation and defibrillation, and these applications are discussed in detail below. These studies have used a range of simplified and biophysically detailed cell models, and both mon-odomain and bidomain tissue models.

5.3 *Whole-organ mechanics models*

Ventricular mechanics models have developed from axisymmetric geometries using thin-walled ellipsoids (Sandler and Dodge, 1963), to isotropic models with realistic longitudinal cross-sections (Gould *et al.*, 1972), and to 3-D ventricular models with transversely isotropic mechanical properties (Hunter *et al.*, 1975). Such models have been useful for understanding the gross mechanical function of the LV in terms of pressure, ejection fraction, and changes in overall dimension, in particular along the apex-to-base axis. Quantification of the fibre-sheet anisotropy for the whole ventricles (LeGrice *et al.*, 1995; Nielsen *et al.*, 1991) led to the development of anatomically realistic ventricular mechanics models (Kerchoffs *et al.*, 2006; Nash and Hunter, 2000; Usyk *et al.*, 2002), which were used to investigate regional deformations of the ventricular walls such as the torsional deformations that facilitate systolic ejection together with the role of tissue anisotropy in this process. This type of modelling has also been used to better understand the effects of disease states such as myocardial infarction on the mechanical performance of the heart (Remme *et al.*, 2005).

5.4 *Coupled electromechanical models*

Many whole-heart mechanical models have incorporated some sort of contraction sequence in order to mimic systolic contraction. Typically, these models incorporated one-way (weak) coupling of an electrical model or a prescribed excitation sequence (which provided the spatio-temporal dynamics of intracellular calcium or fibre tension) in order to mimic the excitation-contraction process (Nash and Hunter, 2000). Other weakly coupled models have been used to analyse the mechano-electrical feedback effects of wall deformations on the tissue electrophysiology and hence wave propagation (Vetter and McCulloch, 2001). To date one of the most comprehensive models of cardiac electromechanics incorporated strongly (2-way) coupled excitation-contraction and mechano-electrical feedback mechanisms in an orthotropic, ellipsoidal LV geometry (Nickerson *et al.*, 2006). This type of modelling has several potential clinical applications, such as the determination of optimal lead placement for bi-ventricular pacemakers in order to maximize cardiac performance. One drawback is that these models presently require the use of high-performance computers because of the computationally intensive nature of their iterative electromechanical algorithms.

6 Numerical and computational issues

Obtaining solutions for the whole-organ models described in the previous section is difficult. There are challenges in the numerical mathematics required to deal with the coupled systems of equations, in implementing these equations as simulation code, and in locating appropriate computational resources. We discuss each of these challenges briefly below.

A simplified representation of a coupled electromechanical model is shown in *Figure 3*. These different components can be solved using different numerical approaches. The electrical models are typically systems of stiff and non-linear ordinary differential equations, and these may be solved using a range of implicit or explicit techniques that are described in standard texts (Press *et al.*, 1992). For biophysically detailed electrophysiology models a very short time step of less than 10 µs may be required, although adaptive approaches can be used so that a short time step is only used when it is necessary. The tissue model may be solved using approaches based on finite differences (Clayton and Panfilov, 2008), finite elements (Vigmond *et al.*, 2002), or finite volumes (Trew *et al.*, 2005). Models of mechanics are generally solved using a finite element approach (Oden, 1972), and typically require a lower spatial and temporal resolution than electrical tissue models.

Implementing numerical solutions as code is a demanding task, although there is a trend towards making code available as open source. Transcribing cell model equations from a published paper into code has been recognized as a major problem, in part because of undetected errors in the typesetting of equations, and errors in the published values of parameters and initial conditions for the model equations. Source code for some models has been made available as an online resource (see for example http://www.case.edu/med/CBRTC/LRdOnline/). However, a more generic approach to this problem has been the development of CellML, which is an XML-based language for encoding cell models (http://www.cellml.org/). Many of the more widely used cardiac cell models are available as CellML, and open source parsers to translate CellML

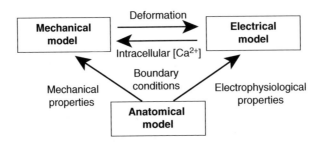

Figure 3. *Interactions between components of a coupled electromechanical model of the heart.*

into source code are in an advanced stage of development. Simulation environments for electrical tissue and mechanics models are also available for academic use. The UCSD Continuity package (http://www.continuity.ucsd.edu/) and the University of Auckland's CMISS package (http://www.cmiss.org/) are biological simulation tools that can be freely downloaded for academic purposes.

Whole-organ coupled electromechanics simulations are significant and expensive computational tasks, and at the time of writing this chapter these tasks need specialized high-performance computational resources with a large number of CPUs able to operate in parallel. However, the development of multi-core CPUs is a recent trend in hardware design, and it is likely that in the medium term high-performance computing will become much more readily available. In addition, computational demands can be reduced using code profiling tools and there is the potential for greater efficiency and faster run-times with many of the existing large-scale heart models.

7 Two success stories

In this chapter we have so far surveyed the overall approach to modelling the heart, as well as covering the physiological background and the types of models that have been developed and used. There are many active research areas where modelling is proving to be a valuable tool, and in this section we have selected two specific areas related to understanding the mechanisms of cardiac arrhythmia. It is well recognized that heart disease is a major cause of premature death in the industrialized world, and in many cases death results from a cardiac arrhythmia called ventricular fibrillation (VF). VF is characterized by a state of electrical anarchy, where instead of being subject to regular activity originating in the sinus node, electrical activity in the ventricles circulates in a self-sustaining way. The mechanisms that initiate and sustain VF are not well understood. A normal rhythm can be restored by prompt application of an electric current, but the reasons why electrical defibrillation sometimes fails are not well understood in detail.

These two success stories focus on events at molecular, tissue, and organ scale. In the modelling studies described, the model has been designed to address a specific research question, and makes appropriate simplifications. For example, none of the models described include mechanics.

7.1 Ion channel phenotypes of gene polymorphisms

In the mid 1990s, advances in molecular biology identified the molecular phenotypes of several gene polymorphisms associated with a high incidence of cardiac arrhythmias (Priori et al., 1999a, 1999b). Most prominent in these studies were genes associated with the Long QT syndrome (LQTS), where ventricular repolarization is prolonged resulting in a lengthened QT interval in the ECG. Mutations of the SCN5A gene were found to be associated with the LQTS, and SCN5A was found to code for the cardiac Na^+ ion channel. One of these, the ΔKPQ mutation, which is a deletion of three amino acids, was found to be associated with altered inactivation of the cardiac Na^+ channel.

The key question arising from these painstaking molecular level studies was precisely how these molecular phenotypes affected cell- and tissue-level behaviour in the heart. This question was answered using a model of normal and mutant cells, which was able to link experimental data from single ion channels with the tissue and organ level LQTS phenotype (Clancy and Rudy, 1999). This was not straightforward, because the Hodgkin–Huxley formulation described above, and used in most cardiac cell models, was not able to reproduce the state-specific molecular phenotype of SCN5A mutations. A new Markovian model of Na^+ channel kinetics was therefore developed for both normal and mutant channels, with a set of states corresponding to open, closed, and inactivated channels. The state transitions and transition rates in the models were developed from single channel experimental data. When this novel Na^+ channel model was incorporated into a model of ventricular cells, the simulated ΔKPQ mutation on the whole-cell action potential not only prolonged the action potential, but also reproduced arrhythmogenic repolarization abnormalities at slow heart rates.

This overall approach of using a model to link molecular and whole-cell behaviours has been successfully applied to other gene mutations underlying the LQTS (Clancy and Rudy, 2001), and as genetic influences on normal and abnormal cardiac function become better understood, modelling is becoming established as a powerful and insightful tool that is capable of linking molecular phenotypes to whole cell behaviour.

7.2 *Models of fibrillation*

Ventricular fibrillation is a lethal arrhythmia, yet is difficult to study experimentally because recordings of electrical activation are limited to the tissue surface. Despite this limitation, there is good evidence to indicate that re-entry is an important mechanism for sustaining VF. During re-entry an action potential propagates continually into recovered tissue, forming a spiral shaped wave of activation. During fibrillation, multiple re-entrant waves propagate throughout the tissue. A 2-D simulation of a single re-entrant wave is shown in *Figure 4A*, and a 2-D simulation of multiple re-entrant waves in *Figure 4B* (see colour plate section). The notion of multiple re-entrant waves as a mechanism for sustaining fibrillation was explored using one of the earliest models of cardiac tissue based on a cellular automaton (Moe *et al.*, 1964), and more recent studies have focused on the mechanism by which a single re-entrant wave can become unstable and break down into multiple wavelets (Fenton *et al.*, 2002).

A 2-D spiral wave rotates around a point, which represents a wavebreak that acts as a source for the continuing activity. In three dimensions, the activation wavefront adopts a scroll shape and rotates around a linear filament. In the ventricles filaments act as the engine that sustains VF, but it is only possible to observe them where they intersect the surface. The motion and interaction of filaments may hold important clues to understanding VF, and this is an active area of research (Clayton *et al.*, 2006). *Figure 4C* shows a snapshot of a 3-D simulation of multiple interacting re-entrant waves. The simulated tissue slab represents a piece of the ventricular wall, and incorporates fibre rotation. In this image, the depolarized regions of tissue are enclosed by coloured isosurfaces, and spirals similar to those shown in *Figure 4B* can be seen on the surface of the simulated tissue. *Figure 4D* shows a simultaneous snapshot of the filaments in this simulation. Several different filament configurations can be seen, and a movie of this simulation would show filaments writhing, merging, splitting in two, appearing out of nowhere, and disappearing. Simulations such as this can be used to count the number of filaments, to estimate the proportion of filaments visible on the surface, and to investigate filament motion and interactions. This information enables experimental observations of tissue surface activity and the electrocardiogram during VF to be interpreted in terms of activation patterns within the ventricular wall.

   This type of approach links cell, tissue, and whole-organ phenomena, and simulations of VF have been undertaken using both simplified and biophysically detailed cell models, and abstract as well as anatomically detailed geometries (Clayton *et al.*, 2006). *Figures 4E–G* show snapshots of simulated VF using a simplified cell model in a geometrical model of canine anatomy.

8 Concluding remarks

The increased availability of high-performance computational resources, coupled with ever more detailed experimental information can be expected to lead to more sophisticated models that are able to answer more complex research questions. There is a trend for more complexity at all levels. Electromechanical models of the heart are beginning to be coupled to models of fluid dynamics and blood flow, and fully working models of the beating heart are close to completion and will enable detailed studies of cardiac haemodynamics. At the molecular and cell scale, models of metabolism will enable comprehensive study of cardiac energetics, as well as the impact of ischaemia.

   In this chapter we have reviewed models of the mammalian heart. These models have already proved their worth as investigative tools and are set to make ever more important contributions to our knowledge of how the heart works in health and disease. They are a real success story in systems biology, because they provide a working illustration of how 'middle out' multi-scale modelling can be used, in partnership with experimental work, to provide completely new insights into the behaviour of complex physiological systems.

References

Augenstein, K.F., Cowan, B.R., LeGrice, I.J., Nielsen, P.M. & Young, A.A. (2005) Method and apparatus for soft tissue material parameter estimation using tissue tagged Magnetic Resonance Imaging. *J. Biomech. Eng.* **127**: 148–157.

Beeler, G.W. & Reuter, H. (1977) Reconstruction of the action potential of ventricular myocardial fibres. *J. Physiol.* **268**: 177–210.

Bers, D.M. (2002) Cardiac excitation-contraction coupling. *Nature* **415**: 198–205.

Burnes, J.E., Ghanem, R.N., Waldo, A.L. & Rudy, Y. (2001) Imaging dispersion of myocardial repolarization, I – Comparison of body-surface and epicardial measures. *Circulation* **104**: 1299–1305.

Burton, R.A.B., Plank, G., Schneider, J.E. *et al.* (2006) 3-Dimensional models of individual cardiac histo-anatomy: tools and challenges. *Ann. NY Acad. Sci.* **1380**: 301–319.

Clancy, C.E. & Rudy, Y. (1999) Linking a genetic defect to its cellular phenotype in a cardiac arrhythmia. *Nature* **400**: 566–569.

Clancy, C.E. & Rudy, Y. (2001) Cellular consequences of HERG mutations in the long QT syndrome: precursors to sudden cardiac death. *Cardiovascular Res.* **50:** 301–313.

Clayton, R.H. & Panfilov, A.V. (2008) A guide to modelling cardiac electrical activity in anatomically detailed ventricles. *Prog. Biophys. Mol. Biol.* **96:** 19–43.

Clayton, R.H., Zhuchkova, E.A. & Panfilov, A. (2006) Phase singularities and filaments: Simplifying complexity in computational models of ventricular fibrillation. *Prog. Biophys. Mol. Biol.* **90:** 378–398.

Costa, K.D., Holmes, J.W. & McCulloch, A.D. (2001) Modelling cardiac mechanical properties in three dimensions. *Phil. Trans. R. Soc. A* **359:** 1233–1250.

Crampin, E., Halstead, M., Hunter, P., Nielsen, P., Noble, D., Smith, N. & Tawhai, M. (2004) Computational physiology and the physiome project. *Exp. Physiol.* **89:** 1–26.

DiFrancesco, D. & Noble, D. (1985) A model of cardiac electrical activity incorporating ionic pumps and concentration changes. *Phil. Trans. R. Soc. London B* **307:** 353–398.

Dobrzynski, H., Li, J., Tellez, J. *et al.* (2005) Computer three-dimensional reconstruction of the sinoatrial node. *Circulation* **111:** 846–854.

Dokos, S., Smaill, B.H., Young, A.A. & LeGrice, I.J. (2002) Shear properties of passive ventricular myocardium. *Am. J. Physiol. (Heart & Circulatory Physiol.)* **283:** H2650–H2659.

Fenton, F.H., Cherry, E.M., Hastings, H.M. & Evans, S.J. (2002) Multiple mechanisms of spiral wave breakup in a model of cardiac electrical activity. *Chaos* **12:** 852–892.

Fox, J.J., McHarg, J.L. & Gilmour, R.F. (2002) Ionic mechanism of electrical alternans. *Am. J. Physiol. (Heart & Circulatory Physiol.)* **282:** H516–H530.

Gould, P., Ghista, D., Brombolich, L. & Mirsky, I. (1972) In vivo stresses in the human left ventricular wall: analysis accounting for the irregular 3-dimensional geometry and comparison with idealised geometry analyses. *J. Biomechanics* **5:** 521–539.

Guccione, J.M., McCulloch, A.D. & Waldmann, L.K. (1991) Passive material properties of intact ventricular myocardium determined from a cylindrical model. *J. Biomech. Eng.* **113:** 42–55.

Guccione, J.M., Waldmann, L.K. & McCulloch, A.D. (1993) Mechanics of active contraction in cardiac muscle: Part II – Cylindrical models of the systolic left ventricle. *J. Biomech. Eng.* **115:** 82–90.

Helm, P., Faisal, M., Miller, M.I. & Winslow, R.L. (2005) Measuring and mapping cardiac fiber and laminar architecture using diffusion tensor imaging. *Ann. NY Acad. Sci.* **1047:** 296.

Hodgkin, A.L. & Huxley, A.F. (1952) A quantitative description of membrane current and its application to conduction and excitation in nerve. *J. Physiol. (London)* **117:** 500–544.

Humphrey, J.D., Strumpf, R.K. & Yin, F.C. (1990a) Determination of a constitutive relation for passive myocardium: I. A new functional form. *J. Biomech. Eng.* **112:** 333–339.

Humphrey, J.D., Strumpf, R.K. & Yin, F.C. (1990b) Determination of a constitutive relation for passive myocardium: II. Parameter estimation. *J. Biomech. Eng.* **112:** 340–346.

Hunter, P.J., McNaughton, P.A. & Noble, D. (1975) Analytical models of propagation in excitable cells. *Prog. Biophys. Mol. Biol.* **30**: 99–144.

Hunter, P.J., McCulloch, A.D. & ter Keurs, H.E. (1998) Modelling the mechanical properties of cardiac muscle. *Prog. Biophys. Mol. Biol.* **69**: 298–331.

Huxley, A.F. (1957) Muscle structure and theories of contraction. *Prog. Biophys. Biophys. Chem.* **7**: 255–318.

Huxley, A.F. & Simmons, R.M. (1971) Proposed mechanism of force generation in striated muscle. *Nature* **233**: 533–538.

Kerchoffs, R.C.P., Healy, S.N., Usyk, T.P. & McCulloch, A.D. (2006) Computational methods for modeling cardiac electromechanics. *Proc. IEEE* **94**: 769–783.

Kleber, A.G. & Rudy, Y. (2004) Basic mechanisms of cardiac impulse propagation and associated arrhythmias. *Physiol. Rev.* **84**: 431–488.

Kohl, P. & Sachs, F. (2001) Mechanoelectrical feedback in cardiac cells. *Phil. Trans. Royal Soc. A* **359**: 1173–1185.

Le Bihan, D., Mangin, J.-F., Poupon, C., Clark, C.A., Pappata, S., Molko, N. & Chabriat, H. (2001) Diffusion tensor imaging: Concepts and applications. *J. Mag. Res. Imaging* **13**: 534–546.

LeGrice, I.J., Smaill, B.H., Chai, L.Z., Edgar, S.G., Gavin, J.B. & Hunter, P.J. (1995) Laminar structure of the heart: ventricular myocyte arrangement and connective tissue architecture in the dog. *Am. J. Physiol. (Heart & Circulatory Physiol.)* **269**: 571–582.

Luo, C.-H. & Rudy, Y. (1991) A model of the ventricular cardiac action potential. Depolarization, repolarization and their interaction. *Circulation* **68**: 1501–1526.

Luo, C.-H. & Rudy, Y. (1994) A dynamic-model of the cardiac ventricular action-potential.1. Simulations of ionic currents and concentration changes. *Circulation Res.* **74**: 1071–1096.

Malvern, L.E. (1969) *Introduction to the Mechanics of a Continuous Medium.* Prentice-Hall, Englewood Cliffs, NJ.

Moe, G.K., Rheinboldt, W.C. & Abildskov, J.A. (1964) A computer model of atrial fibrillation. *Am. Heart J.* **67**: 200–220.

Nash, M.P. & Hunter, P.J. (2000) Computational mechanics of the heart. *J. Elasticity* **61** (1–3): 112–141.

Nickerson, D.P., Nash, M.P., Nielsen, P.M., Smith, N.P. & Hunter, P.J. (2006) Computational multiscale modeling in the physiome project: modeling cardiac electromechanics. *IBM J. Res. Dev.* **50**: 617–630.

Niederer, A.A., Hunter, P.J. & Smith, N.P. (2006) A quantitative analysis of cardiac myocyte relaxation: a simulation study. *Biophys. J.* **90**: 1697–1722.

Nielsen, P.M., LeGrice, I.J., Smaill, B.H. & Hunter, P.J. (1991) Mathematical model of geometry and fibrous structure of the heart. *Am. J. Physiol. (Heart & Circulatory Physiol.)* **260**: 1365–1378.

Noble, D. (1962) A modification of the Hodgkin–Huxley equations applicable to purkinje fibre action and pace-maker potentials. *J. Physiol.* **160**: 317–352.

Noble, D. (2006) *The Music of Life. Biology Beyond the Genome.* Oxford University Press, Oxford.

Noble, D. & Rudy, Y. (2001) Models of cardiac ventricular action potentials: iterative interaction between experiment and simulation. *Phil. Trans. R. Soc. London Series A.* **359**: 1127–1142.

Noble, D., Varghese, A., Kohl, P. & Noble, P. (1998) Improved guinea pig ventricular cell model incorporating a dyadic space, IKr and IKs, and length and tension dependent processes. *Canadian J. Cardiol.* **14:** 123–134.

Oden, J.T. (1972) *Finite Elements of Nonlinear Continua.* McGraw-Hill, New York.

Pandit, S.V., Clark, R.B., Giles, W.R. & Demir, S.S. (2001) A mathematical model of action potential heterogeneity in adult rat left ventricular myocytes. *Biophys. J.* **81:** 3029–3051.

Plonsey, R. & Barr, R.C. (2000) *Bioelectricity. A Quantitative Approach.* Kluwer Academic, New York.

Press, W.H., Flannery, B.P., Teukolsky, S.A. & Vetterling, W.T. (1992) *Numerical Recipes in C. The Art of Scientific Computing.* Cambridge University Press, Cambridge.

Priori, S.G., Barhanin, J., Hauer, R.N.W. *et al.* (1999a) Genetic and molecular basis of cardiac arrhythmias: Impact on clinical management – Part III. *Circulation* **99:** 674–681.

Priori, S.G., Barhanin, J., Hauer, R.N.W. *et al.* (1999b) Genetic and molecular basis of cardiac arrhythmias: Impact on clinical management – Parts I and II. *Circulation* **99:** 518–528.

Puglisi, J.L. & Bers, D.M. (2001) LabHEART: an interactive computer model of rabbit ventricular myocyte ion channels and Ca transport. *Am. J. Physiol. (Heart & Circulatory Physiol.)* **281:** C2049–C2060.

Remme, E.W., Nash, M.P. & Hunter, P.J. (2005) Distributions of myocyte stretch, stress, and work in models of normal and infarcted ventricles. In: *Cardiac Mechano-Electric Feedback and Arrhythmias: From Pipette to Patient* (eds M.R. Franz, P. Kohl and F. Sachs). Saunders, UK, pp. 381–391.

Rudy, Y. (2000) From genome to physiome: Integrative models of cardiac excitation. *Ann. Biomed. Eng.* **28:** 945–950.

Rudy, Y. (2006) Computational biology in the study of cardiac ion channels and cell electrophysiology. *Q. Rev. Biophys.* **39:** 57–116.

Sandler, H. & Dodge, H.T. (1963) Left ventricular tension and stress in man. *Circulation Res.* **13:** 91–104.

Schmid, H., O'Callaghan, P., Nash, M.P., Lin, W., LeGrice, I.J., Smaill, B.H., Young, A.A. & Hunter, P.J. (2008) Myocardial material parameter estimation – a non-homogeneous finite element study from simple shear tests. *Biomech. Model. Mechan.* (in press).

Smith, D.A. (1990) The theory of sliding filament models for muscle contraction. III. Dynamics of the five state model. *J. Theoret. Biol.* **146:** 433–436.

Stevens, C., Remme, E., LeGrice, I.J. & Hunter, P.J. (2003) Ventricular mechanics in diastole: Material parameter sensitivity. *J. Biomechanics* **36:** 737–748.

TenTusscher, K.H.W.J., Noble, D., Noble, P.J. & Panfilov, A.V. (2004) A model for human ventricular tissue. *Am. J. Physiol. (Heart & Circulatory Physiol.)* **286:** H1573–H1589.

Trew, M., LeGrice, I., Smaill, B. & Pullan, A. (2005) A finite volume method for modeling discontinuous electrical activation in cardiac tissue. *Ann. Biomed. Eng.* **33:** 590–602.

Usyk, T.P., LeGrice, I.J. & McCulloch, A.D. (2002) Computational model of three-dimensional cardiac electromechanics. *Comput. Vis. Sci.* **4:** 249–257.

Vetter, F.J. & McCulloch, A.D. (1998) Three dimensional analysis of regional cardiac function: A model of rabbit ventricular anatomy. *Prog. Biophys. Mol. Biol.* **69:** 157–183.

Vetter, F.J. & McCulloch, A.D. (2001) Mechanoelectrical feedback in a model of the passively inflated left ventricle. *Ann. Biomed. Eng.* **29:** 414–426.

Vigmond, E., Aguel, F. & Trayanova, N. (2002) Computational techniques for solving the bidomain equations in three dimensions. *IEEE Trans. Biomed. Eng.* **49:** 1260–1269.

Wei, D. (1997) Whole heart modelling: Progress, principles, and applications. *Prog. Biophys. Mol. Biol.* **67:** 17–66.

Zahalak, G.I. (2000) The two-state cross-bridge model of muscle is an asymptotic limit of multi-state models. *J. Theoret. Biol.* **204:** 67–82.

10

Modelling root growth and development

Eric M. Kramer, Xavier Draye and Malcolm J. Bennett

1 Introduction

Modern biological research largely focuses on the generation of large data sets describing the state of a cell or tissue in terms of patterns of gene activation (transcriptomics), protein content (proteomics), and metabolic state (metabolomics, ionomics). The data generated by 'omics' techniques is generally contained in large computer databases that can be integrated into computer models of gene or other networks employing systems biology approaches (see other chapters in this book). By comparison, the use of systems biology approaches to connect such data to larger scale processes – happening at the tissue, organ, organism, or even ecosystem scale – is relatively unexplored. The reason, of course, is the infancy of the systems biology approach and the daunting practical challenges posed by computer simulations that bridge these multiple physical and temporal scales. However, we anticipate that the next disciplinary frontier will be the construction of large-scale, dynamic computer models that allow an integration of biological data sets obtained from across these physical and temporal scales.

Agriculture, ecology and environmental science have used empirical, phenomenological models of organisms for decades (Thornley and Johnson, 2000). These models are often refined through the incorporation of experimental data, but there continues to be a limited ability to connect the phenotypes of interest to the underlying molecular networks contained within each cell. The construction of a truly multi-scale model aims to finally connect genotype and phenotype in a *quantitative* and *predictive* manner. This is the focus of 'integrative biology'. Integrative biology is so called because, in addition to the essential results of systems biology, it requires an equally thorough knowledge of the relevant biochemistry, biophysics, cell biology, physiology, ecophysiology and the interactions of an organism with its environment. It therefore attempts to integrate results across the entire range of biologically relevant disciplines. This goal is very ambitious and may take decades to fully realize.

In this chapter, we review some recent computational advances in the development and growth of plant roots. In Section 2, we review the current state of phenomenological models of root growth and development, where 'functional-structural models' are the current state of the art in agriculture and environmental science. In Section 3, we review recent computer models of plant hormone transport in roots, with a focus on the hormone auxin. In Section 4, we discuss the development of more sophisticated root models anticipated in the near future. Among other challenges, these will have to incorporate other signals besides auxin.

2 Background to modelling root development

During the last two decades, plant roots have received increasing attention from plant biologists and agronomists, motivated by future agricultural challenges such as sustainable or low-input farming (among many others). Nutrition and anchorage are the major functions of root systems in many plant species. These functions involve complex interactions between the plant and the soil environment and depend quantitatively on root architecture. For example, the availability of water and nutrients varies extensively within the soil profile and during the growing season (Robinson, 1994), so that matching the spatial distribution of roots with those of soil resources may ultimately be as important to plant nutrition as root activity and uptake kinetics (Walk *et al.*, 2006). Similarly, plant anchorage depends on an appropriate balance between the distribution of roots and the (spatially variable) physical properties of the surrounding soil (Dupuy *et al.*, 2005). Plant root systems can be productively analysed as large and dynamic populations of objects (i.e. root segments and meristems) that interact with each other, with shoot organs and with their environment in typically non-linear ways (Walch Liu *et al.*, 2006). Models of root growth and development are thus likely to yield interesting insights into root development and functions, at various scales and in a multidisciplinary framework.

A wide variety of root system models have been developed since the early work of Hackett and Rose (1972) and Lungley (1973). Many of them were introduced as part of a larger model system (plant, soil-plant and soil-plant-atmosphere systems) in which roots were involved as an uptake system or as a sink for photo-assimilates (reviewed by Pagès, 2002). These models, which had the immediate goal of predicting crop functioning or yield or examining ecological issues in the soil-plant-atmosphere system, relied on major simplifications of root morphology and functioning which were specifically designed for the ease of calculations and in order to match data availability. These simplifications were justified in view of the objectives, yet they were hiding much of the complexity of root development and function.

Beginning in the early 1990s, a class of root models was introduced that coupled root developmental dynamics and geometry (*Figure 1*; Lynch *et al.*, 1997; Pagès *et al.*, 2004). Relying on more than a century of macroscopic observation of roots and their soil environment, these models use a set of elementary developmental rules to reproduce in three dimensions the dynamic and autonomous morphogenetic behaviour of individual roots (e.g. root and lateral root formation, elongation, tropisms and decay) and are capable of taking into account spatially variable environmental cues. The separation of these morphogenetic processes is necessary because they operate in the plant at different places and times and involve specific interactions with the environment. By comparing model output with observed root morphology, these models provide a link between a complex object usually observed as static (i.e. root morphology) and the underlying morphogenetic processes.

Building on these architectural models, so-called functional structural plant models (FSPMs) have been developed to explore the complex interactions between plant architecture and the physical and biological processes that drive plant development (Godin and Sinoquet, 2005). For example, the merging of a root architecture model with a soil water-transport (Doussan *et al.*, 1998) or nutrient-transport model

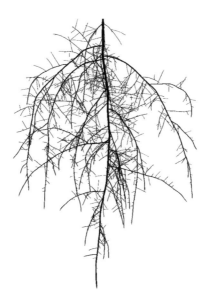

Figure 1. *Simulation of a primary root system of* Arabidopsis thaliana *generated with the root system architecture model from Pagès et al. (2004).*

(Dunbabin *et al.*, 2002; Somma *et al.*, 1998) enables the behaviour of individual (virtual) roots to depend on local environmental conditions. This, in turn, allows for the dynamic simulation of many spatially distributed system variables which may influence subsequent root system growth, and for the simulation of root activity from the scale of the soil-root interface to the whole plant. This approach has also been extended by considering the endogenous environment of each root segment, thus allowing for the representation of interactions between different organs within the plant (Pagès, 2000). A typical example models the competition between parts of the plant for photo-assimilates, the distribution of sugar throughout the root system, and the subsequent influence on the elongation rate of individual roots (Bidel *et al.*, 2000; Thaler and Pagès, 1998).

FSPMs provide a conceptual framework to address many types of interactions at the appropriate scale. However, they are typically concerned with length scales large compared to individual plant cells, mostly from the organ to the plant, but also from tissue to organ or from plant to stand (Godin and Sinoquet, 2005). The necessity of dealing with a large number of plant constituents and the potentially large numbers of interacting processes would make it difficult to integrate details of processes operating at the molecular and cellular scale. Developing models that incorporate knowledge from the molecular to the organ scale is the object of integrative systems biology (ISB) and is extensively discussed below. Eventually, FSPMs and ISB are meant to be connected. Indeed, individual plant organs are affected by internal and external conditions, which requires a reference to their 3-D location in the plant and in the soil, while at the same time, organs contribute to the flow of resources or regulatory signals delivered within the root system network (Birnbaum and Benfey, 2004; Forde, 2002; Malamy, 2005).

3 Modelling hormone regulated root development

Integrative systems biology (ISB) promises to revolutionize the way root growth and development will be modelled in the future. ISB approaches will enable researchers to construct models of developmental processes that incorporate molecular, cellular and tissue scale information. Recent advances in our knowledge about the molecular transport and signal transduction of the hormone auxin (reviewed by Kramer and Bennett, 2006; Leyser, 2006; Teale, 2006) have made it the signal molecule of choice for many conceptual models of hormone-mediated development. Recent models have examined the role of auxin in phyllotaxis (de Reuille *et al.*, 2006; Heisler and Jonsson, 2006; Jonsson *et al.*, 2006; Smith *et al.*, 2006), wood grain pattern formation (Forest *et al.*, 2006; Kramer, 2002, 2006; Kramer and Borkowski, 2004; Kramer and Groves, 2003), root gravitropism (Swarup *et al.*, 2005), and the development of vascular strands in stems and leaves (Dimitrov and Zucker, 2006; Feugier and Iwasa, 2006; Feugier *et al.*, 2005; Kramer, 2004; Mitchison, 1980b, 1981; Rolland-Lagan and Prusinkiewicz, 2005; Runions *et al.*, 2005). The primary motivation for most of these papers is (i) to predict the movement of auxin within a developing tissue and (ii) to show that some hypothesis for auxin's role as a developmental signal, implemented *in silico*, gives a simulation output comparable to the observed patterns *in vivo*. To date, most of these models do not include even simplified gene networks, but focus only on one or more auxin transport components. There have also been few systematic attempts to refine model hypotheses using new experiments – a key feature of the ISB approach.

In the following sections we will focus on models of auxin transport, but wish to emphasize our expectation that a complete model of plant development will eventually need to incorporate many other hormonal, nutritional and environmental signals within the tissue.

3.1 *A primer on computer models for auxin transport*

Auxin transport is trans-cellular, which is to say it moves from the cytoplasm of one cell to the next by crossing both cell membranes and the wall that separates them. There are three gene families functionally associated with auxin transport across the cell membrane AUX/LAX in influx (Parry *et al.*, 2001; Yang *et al.*, 2006), PIN in efflux (Benkova *et al.*, 2003; Petrasek *et al.*, 2006), and MDR/PGP in both (Geisler *et al.*, 2005; Terasaka *et al.*, 2005). Auxin is also weakly membrane permeable when protonated (Kramer and Bennett, 2006). Thus, a complete model of auxin transport may include (i) the diffusion of auxin within the cytoplasm and/or within the extracellular space (the apoplast), (ii) the partition of auxin between cytoplasm, vacuole, and other relevant sub-cellular compartments, and (iii) the spatial localization of carriers on the cell membrane and the resulting auxin fluxes. The direction of auxin transport is determined by the asymmetric localization of auxin efflux, and sometimes auxin influx, carriers at the plasma membrane surface of plant cells. This is illustrated in *Figure 2* (see colour plate section) where the asymmetric localization of the auxin efflux carrier protein PIN2 in lateral root cap and epidermal cells result in auxin flowing away from the root apex (Muller *et al.*, 1998).

Discretization. Models of auxin transport describe a plant tissue as a collection of polygonal or polyhedral boxes (in two or three dimensions, respectively). There are many synonyms for 'box' in the literature, including lattice site, volume element, control volume, and cell. The technique of describing an object as a set of boxes to permit computer simulation is called *discretization*. The size and arrangement of boxes is not just a technical convenience. Rather, the discretization reflects many implicit and explicit assumptions about the relevant cell and molecular biology to be modelled.

The most spatially detailed models published to date subdivide the cytoplasm and apoplast into enough boxes to resolve the sub-cellular auxin gradient and the U-shaped distribution of efflux carriers on the cell membrane (Kramer, 2004; Swarup et al., 2005; *Figure 3* (see colour plate section)). However, models with this level of resolution are rare.

The most common choice for the discretization of a plant tissue is an assignment of exactly one box per plant cell (de Reuille *et al.*, 2006; Feugier and Iwasa, 2006; Feugier *et al.*, 2005; Jonsson *et al.*, 2006; Mitchison, 1980b, 1981; Rolland-Lagan and Prusinkiewicz, 2005; Smith *et al.*, 2006). We will call these 'cell models'. A cell model cannot describe the gradient of auxin within the cell. Nor can it resolve the distribution of auxin within different sub-cellular compartments (vacuole, cytoplasm, nucleus, etc). Lastly, cell models assume that the diffusion of auxin in the extracellular space does not contribute significantly to the overall flux through the tissue. The use of a cell model makes the implicit assumption that such details are not relevant for the topic under investigation. The validity of this assumption is reviewed at length by Kramer (2008).

A third choice for the discretization assigns many plant cells – tens to thousands – to each box (Dimitrov and Zucker, 2006; Douady and Couder, 1992; Kramer, 2002, 2006; Kramer and Borkowski, 2004; Kramer and Groves, 2003; Mitchison, 1977). These may be called 'continuum' or 'coarse-grained' models. The concentration of auxin and other hormones, as well as patterns of gene activation, are averaged over all the cells in the box. The main advantage of this approach is economy. Because individual cells are not represented, whole plant organs can be simulated in relatively short computer time. Additionally, in systems that have been poorly characterized experimentally, phenomenological models can be devised with just a few undetermined parameters. These advantages should be weighed against the fact that biologists usually have little training in continuum approaches, and understandably prefer models that resolve individual cells.

Transport. Computer models of auxin transport all work in much the same way. Each box contains an amount of auxin. The concentration of auxin in each box is repeatedly recalculated according to a set of mathematical rules. One calculation step, applied to the whole model tissue, is called one *iteration*. The core of any model is the algorithm that dictates how the amount of auxin in each box is updated during each iteration. Roughly speaking, the amount of auxin in a box can change due to *de novo* synthesis of auxin, metabolization of auxin, and transport of auxin in or out across the boundary of the box. The algorithm chosen depends in part on the choice of discretization, and in part on the cell biology one wishes to model.

The mathematical description of the fluxes in auxin transport models is based on well-established concepts in biophysics (Benedek and Villars, 2000; Jackson, 2006). Diffusion between a pair of adjacent boxes in the same compartment (e.g. two boxes in the same vacuole) is governed by Fick's law.

$$J_{1\to 2}=-D\frac{c_2-c_1}{L} \tag{1}$$

where $J_{1\to 2}$ is the net flux from box 1 to box 2, c_j is the concentration of auxin in box j (includes both protonated and anion forms), D is the diffusion coefficient of auxin in the compartment, and L is the centre-to-centre distance between the boxes.

The flux of auxin between two boxes separated by a membrane is a sum of terms,

$$J_{1\to 2}=J_{diff}+J_{PIN}+J_{AUX1} \tag{2}$$

where J_{diff}, J_{PIN}, and J_{AUX1} are due to the diffusive membrane permeability of the protonated auxin molecule, PIN family efflux carriers, and AUX family influx carriers, respectively. The diffusive permeability is described by Fick's law and the activity of the efflux carriers by the Goldman-Hodgkin-Katz (GHK) current equation (Jackson, 2006).

$$J_{diff}=-P_{IAAH}\left(\frac{c_2}{1+10^{pH_2-pK}}-\frac{c_1}{1+10^{pH_1-pK}}\right) \tag{3}$$

$$J_{PIN}=-P_{PIN}\left(\frac{f(-\phi)c_2}{1+10^{-pH_2+pK}}-\frac{f(\phi)c_1}{1+10^{-pH_1+pK}}\right) \tag{4}$$

$$f(x)=\frac{x}{e^x-1} \tag{5}$$

where pH_i is the pH of compartment i, pK is the acid-base dissociation constant of auxin, the P_j are membrane permeabilities, and $\phi=\pm FV/RT$ is a normalized potential difference where V is the membrane potential, F is the Faraday constant, R is the gas constant, T is the temperature, and the + sign is used if box 1 is interior to box 2 (by definition, the cytoplasm is interior to the cell wall). In models, the activity of the influx carrier is usually described by the GHK equation as well,

$$J_{AUX}=-P_{AUX}\left(\frac{f(\phi)c_2}{1+10^{-pH_2+pK}}-\frac{f(-\phi)c_1}{1+10^{-pH_1+pK}}\right) \tag{6}$$

although this is only an approximation since the influx carrier involves symport of two protons with each auxin anion (Lomax et al., 1985). The effects of carrier saturation, and pH-dependent or voltage-dependent carrier efficiency can easily be incorporated by treating the permeabilities as functions of pH and ϕ. Future models may include a fourth term in Equation (1) representing the auxin fluxes due to the MDR/PGP family of carriers (Geisler et al., 2005; Noh et al., 2001; Santelia et al., 2005; Terasaka et al., 2005).

Carrier localization. In models intended to capture the dynamic redistribution of auxin efflux carriers, the algorithm must also include rules for moving carriers to and from each membrane segment on the border of the cell (Jonsson *et al.*, 2006; Smith *et al.*, 2006). Since little is known about the signalling mechanism that controls the trafficking of carriers, this portion of the algorithm is phenomenological.

3.2 *Model parameters*

Any algorithm modelling auxin transport requires real numbers as inputs. Parameters may include auxin membrane permeabilities, protein carrier efficiencies, auxin diffusion coefficients, and the rates of auxin synthesis and degradation. Additional parameters may be required to characterize the rates of auxin carrier synthesis, degradation, and redistribution. Ideally, every parameter should be quantified experimentally. However, in many cases one is left with educated guesses based in part on the underlying biophysics and biochemistry, and in part on inference from published values in other systems.

Diffusive permeability. There is a wide range of published values for the permeability of plant lipid membranes to protonated indole-3-acetic acid (IAAH). Many of these must be discounted however, since they do not control for the likely presence of anion carriers (Baier *et al.*, 1990; Gimmler *et al.*, 1981). The only two studies we know of that control for carriers are both by Delbarre and co-workers (Delbarre *et al.*, 1994, 1996). These report permeabilities of 0.14 cm hr^{-1} and 0.18 cm hr^{-1}, respectively, for suspension-cultured tobacco protoplasts. The auxin anion is not membrane permeable (Bean *et al.*, 1968; Gutknecht and Walter, 1980).

Carrier permeabilities. The permeabilities of auxin influx carriers (the AUX/LAX gene family; Parry *et al.*, 2001) and auxin efflux carriers (the PIN gene family; Benkova *et al.*, 2003; Galweiler *et al.*, 1998) are not well known. Perhaps the most useful source of quantitative data is Delbarre *et al.* (1996), who report the diffusive and carrier mediated auxin influx and efflux in suspension-cultured tobacco protoplasts. We estimate an influx carrier permeability of 0.02 cm hr^{-1} and an efflux carrier permeability of 0.01 cm hr^{-1} for their protoplasts. A second useful paper, and the only published value for influx carrier permeability in a plant tissue, is Szponarski *et al.* (1999). They report an influx carrier permeability of 0.011 cm hr^{-1}, measured in plasma membrane vesicles derived from mature *Arabidopsis* leaves (note the surprising agreement with the tobacco protoplast result). Szponarski's result reflects an average over all cells of the leaf, so the permeability in cells specialized for auxin transport is likely to be larger. Considering these facts, carrier permeability values of order 0.1 cm hr^{-1} are expected to be reasonable.

Intracellular diffusion. We do not know of any direct measurements of the cytoplasmic or vacuolar diffusion of IAA (indole-3-acetic acid). Since the vacuole is mostly water, we expect the vacuolar diffusion coefficient of IAA to be comparable to the known aqueous value, $D_{aq} = 0.024$ cm^2 hr^{-1} (Mitchison, 1980a). In the cytoplasm, comparable small molecules have diffusion coefficients in the range of 10% to 40% of their aqueous value (Paine *et al.*, 1975).

Apoplastic diffusion. The permeability of the apoplast to small molecules is not homogeneous, having significant restrictions at Casparian bands, and also possibly within the root meristem (Cholewa and Peterson, 2001; Enstone and Peterson,

1992; Wierzbicka, 1987). Following Richter and Ehwald (1983) and Aloni *et al.* (1998), a reasonable value for the diffusion coefficient of IAA in cell wall material is probably 10% of the aqueous value. The restriction of wall permeability at Casparian bands and other localized hydrophobic barriers may decrease the diffusion coefficient by two or more orders of magnitude (Bayliss *et al.*, 1996; Canny and Huang, 1994).

Robustness. Rather than entering an empirical discussion about parameters, some modelling groups pursue a different approach. Model parameter values are invented with no reference to experiment. These parameter choices are then justified using two arguments. First, model parameters are typically validated *post hoc* by comparison between model output and plant morphology. Second, a range of parameter values are tested to see if the model output is robust under parameter variation (sensitivity analysis). The use of robustness as a selection criterion for parameter values has precedent in studies of biochemical networks (El-Samad *et al.*, 2006; Morohashi *et al.*, 2002). To the extent that this technique succeeds, it is credited to the fact that evolution has tended to select for robustness in biological systems (Kitano, 2004; Li *et al.*, 2004). Robustness analyses may be applied as a useful check on parameters whose values are poorly known. With caution, it may be used to fill in conspicuous gaps in the empirical knowledge of a system. Robustness analysis should not substitute for a careful comparison with experiment in cases where one is available.

3.3 *Employing ISB to probe developmental processes in roots*

The most complete published model of auxin transport in the root appears in Swarup *et al.* (2005). This model concerns auxin transport in the root elongation zone, and is specifically concerned with the role of auxin in root gravitropism. The conventional model of root gravitropism is shown in *Figure 2C* (see colour plate section). Auxin transported to the tip through the stele, and auxin synthesized in the tip, changes direction in the root cap, and moves via the lateral root cap to the elongation zone. During a gravity response (i.e. if the root tip is placed horizontally), basipetal auxin transport favours the accumulation of auxin on the underside of the root cap and the elongation zone. Since auxin inhibits the elongation of cells in the elongation zone, this leads to a downward deflection of the root tip.

Recent work on the localization pattern of auxin carriers in *Arabidopsis* has substantially clarified this picture (*Figure 2*; see colour plate section). First, the auxin efflux carrier PIN3 is expressed in gravity-sensing columella cells at the root tip. Following a gravity stimulus, PIN3 is relocalized to the side of the columella cell facing downwards, causing auxin to be asymmetrically redistributed onto the lower side of the root. Second, the auxin influx carrier AUX1 is expressed in a continuous file of cells that begins at the columella, passes through the lateral root cap, and continues in the elongation zone (*Figure 2A*). This creates a 'canal' for the efficient transport of auxin from the columella, through the lateral root cap, to the elongation zone epidermis. Third, the auxin efflux carrier PIN2 is expressed in the same cells of the canal and localized to a region of the cell membrane that points away from the columella (*Figure 2B*), thereby establishing the basipetal direction of auxin transport within the canal (*Figure 2C*).

The roles of AUX1 and PIN2 were examined using a 3-D model of the elongation zone (*Figure 3*; see colour plate section). As only the epidermis expresses AUX1 in the elongation zone, simulations suggest that this tissue accumulates auxin at a concentration 10–20 times higher than the cortex and endodermis, in agreement with experimental observations (Blilou et al., 2005; Ohwaki and Tsurumi, 1976; Tsurumi and Ohwaki, 1978). For this reason, it is the direction of PIN auxin efflux carriers in the epidermis that establishes the major basipetal direction of flux through the outer root. Flux may be regarded as the product of two terms – a speed (set by the carrier permeability and the cell size) multiplied by a concentration. The relative depletion of auxin from the cortex and endodermis makes them minor players in the auxin flux. Model runs with different choices for the direction of efflux from cortical cells show only small changes in the auxin distribution. The general rule seems to be that PIN efflux carriers establish the direction of transport, whilst the AUX1 influx carrier determines which cell files participate in the transport stream.

Another important consequence of AUX1 being expressed in the epidermis is that it will cause the majority of the lateral auxin gradient to accumulate in the epidermis. Partitioning the lateral auxin gradient in wild-type roots in such a manner would result in root gravitropic bending being driven by differential cell elongation primarily in the epidermis. If this is the case, then selectively disrupting the auxin response in epidermal tissue would disrupt root gravitropism. We tested this model *in planta* by expressing a repressor of the root auxin response termed axr3-1 in selected tissues. Consistent with our model, axr3-1 blocked the auxin-mediated inhibition of root growth when expressed in expanding epidermal cells. Disrupting the auxin response in epidermal tissues resulted in a complete loss of root gravitropic curvature. In contrast, expressing axr3-1 in other tissues had little or no effect on root gravitropism. We therefore conclude that root gravitropic curvature is driven primarily by the differential growth of expanding epidermal cells in response to the lateral auxin gradient. Moreover, these experiments serve to validate the predictions made by our root model.

4 Towards a virtual root model

4.1 *Lessons from above*

To illustrate how models of root development are likely to be constructed in the near future, it is instructive to review recent developments in the modelling of phyllotaxis in the shoot. Phyllotaxis is the name given to the regular spatial and temporal pattern of organ initiation in the apical meristem of stems and reproductive structures. One of the more visually arresting examples of pattern in phyllotaxis is the spiral arrangement of florets in the centre of composite flowers such as sunflower (*Helianthus annuus*). Patterns in spiral phyllotaxis have attracted an enormous amount of attention from mathematicians for more than a century (Jean, 1994). Of particular interest is the fact that the number of left- and right-handed spirals are often two consecutive terms in the Fibonacci series, {1,1,2,3,5,8,13,21...} in which each term is the sum of the previous two.

The reproduction of the Fibonacci series would seem to provide a stringent test for any computer model of development. Surprisingly, this is not the case. Mitchison (1977), and later Douady and Couder (1992), used computer models of the apical meristem to demonstrate that models of spiral phyllotaxis will reproduce the Fibonacci series under a wide range of plausible conditions. These models treat the meristem as a two-dimensional surface (i.e. a flat or curved sheet) and organ primordia as points on this surface. Individual cells are not resolved. The conditions necessary and sufficient to reproduce phyllotactic patterns are the following: (i) new primordia can only arise in an annular region near the tip of the meristem; (ii) existing primordia inhibit the formation of new primordia via a signal whose strength decays with distance; and (iii) once formed, a primordium maintains its identity and is carried away from the tip of the meristem by ongoing tissue growth. The importance of this result for modellers is the fact that it is independent of the nature of the inhibition between primordia. Recently, two distinct theories of this inhibition have both been used to motivate computer models, and both produce realistic phyllotactic patterns. One class of models uses an inhibition based on auxin production and transport and will be discussed further below. The second class of models assumes the pattern arises as a result of the elastic buckling of the relatively stiff outer layer of the meristem as it is subjected to growth stresses (Green, 1999; Shipman and Newell, 2005). The work of Mitchison, Douady and Couder shows that simulated phyllotactic patterns cannot be used to decide between these competing models of development. Instead, the nature of the inhibition must be determined by experiments.

The role of auxin in phyllotaxis has recently been clarified by numerous experiments involving auxin transport inhibitors, auxin carrier mutants, and auxin carrier visualization studies (Galweiler et al., 1998; Reinhardt, 2005; Reinhardt et al., 2000, 2003; Stieger et al., 2002). Both auxin influx and efflux carriers play important roles. In Arabidopsis, the AUX1 influx carrier expression is strongest in the outermost (L1) cell layer of the shoot apical meristem (SAM). The expectation is that this pattern of expression tends to restrict auxin to the outermost layer, thus explaining why primordium initiation is effectively 2-D. The inhibition between neighbouring primordia is explained by a feedback between the auxin concentration and the sub-cellular localization of the PIN1 efflux carrier. The idea is that PIN1 tends to point up the local auxin gradient, towards existing sites of auxin accumulation. Thus, a zone of auxin accumulation will tend to deplete auxin from adjacent cells and to inhibit the formation of nearby concentration maxima. The hypothesis is completed by assuming that localized sites of auxin accumulation differentiate to become organ primordia. Many aspects of this model remain poorly understood. In particular, the signal that conveys to a cell information about the auxin concentration in neighbouring cells is unknown, as is the means by which this signal is conveyed to the mechanism that controls PIN protein trafficking.

The conceptual model of auxin-mediated phyllotaxis was developed without the assistance of mathematical or computer models. Only subsequently, in the last two years, have several groups published computer models of auxin flux in the L1 layer of the Arabidopsis SAM. De Reuille and co-workers (de Reuille et al., 2006) map the distribution of PIN1 obtained from microscopy directly onto a static cell model (one polygonal box per cell, no cell division or growth) and show that the sites of auxin accumulation in silico match the observed sites of primordium initiation. Smith and co-workers (Smith et al., 2006) build a dynamic cell model of the SAM that includes cell division and expansion, auxin transport, and a feedback between the local auxin gradient and efflux carrier localization. They show that the model can reproduce a variety of realistic phyllotactic patterns, differing only in the choice of model transport parameters. Both these cell models are subject to the limitations described in the previous section. In addition, both groups use unitless model parameters. It is there-fore difficult to assess whether their parameter values are consistent with available information on membrane permeability and carrier efficiency (Delbarre *et al.*, 1996; Swarup *et al.*, 2005).

Jonsson *et al.* (2006) apply two different modelling approaches to phyllotaxis. The first is a cell model of the L1 layer including cell division and growth, and using unitless parameters. Like Smith *et al.* (2006), they find that the feedback between PIN polarity and local auxin concentration is sufficient to generate patterns of primordium initiation. Unlike Smith *et al.*, they find that the spatial and temporal pattern of auxin accumulation into discrete zones is unstable, showing reversals of the spiral pattern and uneven time intervals between consecutive primordia initiations. The difference may be due to the fact that Smith's model allows cells at the centre of an auxin accu-mulation zone to undergo an irreversible developmental transition to a primordium, after which the local transport parameters are changed to favour their continued sta-bility. There is no irreversible developmental transition in the model of Jonsson *et al.* The second approach to phyllotaxis described by Jonsson *et al.* (2006) is similar in spirit to that of de Reuille *et al.* (2006). The authors map microscope observations of PIN1 polarity onto a static model of the L1 layer and locate the sites of auxin accu-mulation *in silico*. The model has two notable advantages over those previously described. First, the model explicitly includes the cell wall between each pair of cells. Thus, it can distinguish auxin influx from efflux and also quantify the possible impor-tance of auxin diffusion through the apoplast. Second, all model parameters have units, allowing a comparison with experiment.

In a more recent paper, Heisler and Jonsson (2006) have continued their analysis of this model. They adopt parameter values for both diffusive and carrier-mediated trans-port that have a plausible foundation in experiment. In so doing, they avoid high mem-brane permeabilities that would lead to large cytoplasmic auxin gradients. From the perspective of molecular biology, the chief advance over their previous paper (Jonsson *et al.*, 2006) is the incorporation of the AUX1 influx carrier into the model (recall that cell models historically tend to ignore influx carriers). Two new results emerge from this addition. First, they show that the experimentally observed high level of AUX1 expression in the L1 layer is adequate to accumulate auxin in this layer, supporting ear-lier speculation (Reinhardt *et al.*, 2003). Second, they show that the observed up-regu-lation of AUX1 in cells at the sites of primordium initiation is sufficient to stabilize the position of auxin accumulation sites. This model in particular demonstrates that the integration of realistic biophysical and biochemical details into a computer simulation can illuminate the functional role of genes in plant development.

Although these models are necessarily phenomenological and approximate, as a class they have proven their usefulness in the following ways. They have shown that the conceptual model of phyllotaxis can be implemented plausibly and quantitatively as a set of interactions between cells in the L1 layer. Beyond this, the models highlight several important features that were obscure or missing in the conceptual model. First, the feedback between the local auxin gradient and PIN polarization requires a short-range signal between adjacent cells. Second, the stability and regularity of the phyllotactic pattern depends on the activity of genes other than efflux carriers (either a transition from SAM to primordium identity or an auxin-mediated regulation of influx carrier expression). Third, despite differences in the final choice of differential equations, all three groups identified auxin homeostasis as an important feature of auxin dynamics in the L1 layer.

We anticipate that models of root development in the near future will adopt some of the same techniques as those described above for the shoot: (i) Models of hormone transport will be incorporated into more realistic cell geometries, probably reconstructed from confocal microscope studies. (ii) Some aspects of the signalling network between auxin concentration and gene expression will be incorporated. In particular, the role of auxin in the initiation and growth of new lateral roots is under current investigation (EMK & MJB, unpublished). (iii) The initiation of root primordia in the pericycle has numerous similarities with primordium initiation in the shoot. For example, the early stages of primordium initiation involve a dramatic reorientation of efflux carrier localization in participating cells (Benkova *et al.*, 2003), probably as a result of a feedback mechanism similar to that in the shoot.

4.2 *The need to integrate many more signals*

Whilst recent studies have focused exclusively on modelling auxin flux, genetic studies have revealed that many other classes of hormone signals regulate root growth and development such as brassinosteroids, ethylene and gibberellins (Fu and Harberd, 2003; Müssig *et al.*, 2003; Roman *et al.*, 1995). Additional signals include nutrients collected by the root system from the soil, metabolites transported from the shoot via the phloem, and signals from symbiotic soil organisms (Lambrecht *et al.*, 2000). A virtual root model will therefore require the integration of many signals.

To date, most research into the role of plant hormones has tended to focus on a single player at a time. Consequently, conceptual models of hormone action tend to be linear: hormone synthesis → transport → perception → response. However, there is a competing paradigm that suggests the cell's genetic program takes as inputs all the available signals at the cell surface and within the cytoplasm and 'integrates' these signals into a limited range of outputs, including cell growth and differentiation (Fu and Harberd, 2003). In this view, hormones act in concert like the gears in a machine – the unique function of any one hormone is then an ill-posed question. The true picture probably lies somewhere between these two extremes, with some cell-specific responses controlled by a single hormone and other responses representing a true synthesis of multiple signals. The full topology of the signalling network is under investigation (Nemhauser *et al.*, 2006).

To date, studies in *Arabidopsis* have focused on dissecting the *molecular* basis of signal integration (reviewed in Gazzarrini and McCourt, 2003) revealing, for example, that auxin and gibberellins (GA) control primary root growth by altering the stability of DELLA proteins that function as growth repressors (Fu and Harberd, 2003). Despite such advances in our understanding of the molecular basis of crosstalk, it remains unclear how phytohormone signalling is integrated at the tissue or organ level. Identifying which root tissues perceive each phytohormone represents an essential prerequisite to understanding how these signals collectively control root growth. As a proof of concept, we recently demonstrated that auxin primarily regulates differential root growth via the epidermis (Swarup *et al.*, 2005). Equivalent knowledge for all the other signals listed above is required before we will be able to construct a truly integrated multi-scale model of the root.

Acknowledgements

The authors would like to acknowledge support from the EPSRC and BBSRC CISB programme to CPIB (to E.K. and M.J.B.); Belgian scientific policy (BELSPO, contract BARN to M.J.B. and X.D); ARC program of the Communauté Française de Belgique (to X.D.) and Research Associate grant from the Belgian Fonds de la Recherche Scientifique (to XD).

References

Aloni, R., Enstone, D.E. & Peterson, C.A. (1998) Indirect evidence for bulk water flow in root cortical cell walls of three dicotyledonous species. *Planta* 207: 1–7.

Baier, M., Gimmler, H. & Hartung, W. (1990) The permeability of the guard cell plasma membrane and tonoplast. *J. Exp. Bot.* 41: 351–358.

Bayliss, C., van der Weele, C. & Canny, M.J. (1996) Determination of dye diffusivities in the cell-wall apoplast of roots by a rapid method. *New Phytologist* 134: 1–4.

Bean, R., Shepherd, W. & Chan, H. (1968) Permeability of lipid bilayer membranes to organic solutes. *J. Gen. Physiol.* 52: 495–508.

Benedek, G.B. & Villars, F.M.H. (2000) *Physics with Illustrative Examples from Medicine and Biology: Statistical Physics*, 2nd Edn. AIP Press, New York.

Benkova, E., Michniewicz, M., Sauer, M., Teichmann, T., Seifertova, D., Jurgens, G. & Friml, J. (2003) Local, efflux-dependent auxin gradients as a common module for plant organ formation. *Cell* 115: 591–602.

Bidel, L.P.R., Pagès, L., Riviere, L.M., Pelloux, G. & Lorendeau, J.Y. (2000) MassFlowDyn I: A carbon transport and partitioning model for root system architecture. *Ann. Bot.* 85: 869–886.

Birnbaum, K. & Benfey, P.N. (2004) Network building: transcriptional circuits in the root. *Curr. Opin. Plant Biol.* 7: 582–588.

Blilou I., Xu J., Wildwater M., Willemsen V., Paponov I., Friml J., Heidstra R., Aida M., Palme K. & Scheres B. (2005) The PIN auxin efflux facilitator network controls growth and patterning in Arabidopsis roots. *Nature* 433: 39-44.

Canny, M.J. & Huang, C.X. (1994) Rates of diffusion into roots of maize. *New Phytologist* 126: 11–19.

Cholewa, E. & Peterson, C.A. (2001) Detecting exodermal Casparian bands in vivo and fluid-phase endocytosis in onion (*Allium cepa* L.) roots. *Canadian J. Bot.* 79: 30–37.

Delbarre, A., Muller, P., Imhoff, V., Morgat, J.-L. & Barbier-Brygoo, H. (1994) Uptake, accumulation and metabolism of auxins in tobacco leaf protoplasts. *Planta* 195: 159–167.

Delbarre, A., Muller, P. & Guern, J. (1996) Comparison of mechanisms controlling uptake and accumulation of 2,4-dichlorophenoxy acetic acid, naphthalene-1-acetic acid, and indole-3-acetic acid in suspension-cultured tobacco cells. *Planta* 198: 532–541.

de Reuille, P.B., Bohn-Courseau, I., Ljung, K., Morin, H., Carraro, N., Godin, C. & Traas, J. (2006) Computer simulations reveal properties of the cell-cell signalling network at the shoot apex in *Arabidopsis. Proc. Natl Acad. Sci. USA.* 103: 1627–1632.

Dimitrov, P. & Zucker, S.W. (2006) A constant production hypothesis guides leaf venation patterning. *Proc. Natl Acad. Sci. USA.* **103:** 9363–9368.

Douady, S. & Couder, Y. (1992) Phyllotaxis as a physical self-organized growth process. *Phys. Rev. Lett.* **68:** 2098–2101.

Doussan, C., Pagès, L. & Vercambre, G. (1998) Modelling of the hydraulic architecture of root systems: An integrated approach to water absorption – Model description. *Ann. Bot.* **81:** 213–223.

Dunbabin, V.M., Diggle, A.J., Rengel, Z. & Van Hugten, R. (2002) Modelling the interactions between water and nutrient uptake and root growth. *Plant & Soil* **239:** 19–38.

Dupuy, L., Fourcaud, T. & Stokes, A. (2005) A numerical investigation into the influence of soil type and root architecture on tree anchorage. *Plant & Soil* **278:** 119–134.

El-Samad, H., Prajna, S., Papachristodoulou, A., Doyle, J. & Khammash, M. (2006) Advanced methods and algorithms for biological network analysis. *Proc. IEEE* **94:** 832–853.

Enstone, D. & Peterson, C.A. (1992) The apoplastic permeability of root apices. *Canadian J. Bot.* **70:** 1502–1512.

Feugier, F.G. & Iwasa, Y. (2006) How canalization can make loops: A new model of reticulated leaf vascular pattern formation. *J. Theoret. Biol.* **243:** 235–244.

Feugier, F.G., Mochizuki, A. & Iwasa, Y. (2005) Self-organization of the vascular system in plant leaves: Inter-dependent dynamics of auxin flux and carrier proteins. *J. Theoret. Biol.* **236:** 366–375.

Forde, B.G. (2002) The role of long-distance signalling in plant responses to nitrate and other nutrients. *J. Exp. Bot.* **53:** 39–43.

Forest, L., Padilla, F., Martinez, S., Demongeot, J. & San Martin, J. (2006) Modelling of auxin transport affected by gravity and differential radial growth. *J. Theoret. Biol.* **241:** 241–251.

Fu, X. & Harberd, N.P. (2003) Auxin promotes *Arabidopsis* root growth by modulating GA response. *Nature* **421:** 740.

Galweiler, L., Guan, C., Muller, A., Wisman, E., Mendgen, K., Yephremov, A. & Palme, K. (1998) Regulation of polar auxin transport by AtPIN1 in *Arabidopsis* vascular tissue. *Science* **282:** 2226–2230.

Gazzarrini, S. & McCourt, P. (2003) Cross-talk in plant hormone signalling: What *Arabidopsis* mutants are telling us. *Ann. Bot.* **91:** 605–612.

Geisler, M., Blakeslee, J.J., Bouchard, R. *et al.* (2005) Cellular efflux of auxin catalyzed by the *Arabidopsis* MDR/PGP transporter AtPGP1. *Plant J.* **44:** 179–194.

Gimmler, H., Heilmann, B., Demming, B. & Hartung, W. (1981) The permeability coefficients of the plasmalemma and the chloroplast envelope of spinach mesophyll cells for phytohormones. *Z. Naturforsch.* **36 c:** 672–678.

Godin, C. & Sinoquet, H. (2005) Functional-structural plant modelling. *New Phytologist* **166:** 705–708.

Green, P.B. (1999) Expression of pattern in plants. *Am. J. Bot.* **86:** 1059-1076.

Gutknecht, J. & Walter, A. (1980) Transport of auxin (indoleacetic acid) through lipid bilayer membranes. *J. Membrane Biol.* **56:** 65–72.

Hackett, C. & Rose, D.S. (1972) A model of the extension and branching of a seminal root of barley, and its use in studying relations between root dimensions. *Australian. J. Biol. Sci.* **25:** 669–679.

Heisler, M.G. & Jonsson, H. (2006) Modelin auxin transport and plant development. *J. Plant Growth Regul.* **25**: 302–312.

Jackson, M.B. (2006) *Molecular and Cellular Biophysics.* Cambridge University Press, Cambridge.

Jean, R.V. (1994) *Phyllotaxis: a systemic study in plant morphogenesis.* Cambridge University Press, Cambridge.

Jonsson, H., Heisler, M.G., Shapiro, B.E., Meyerowitz, E.M. & Mjolsness, E. (2006) An auxin-driven polarized transport model for phyllotaxis. *Proc. Natl Acad. Sci. USA.* **103**: 1633–1638.

Kitano, H. (2004) Biological robustness. *Nat. Rev. Genet.* **5**: 826–837.

Kramer, E.M. (2002) A mathematical model of pattern formation in the vascular cambium of trees. *J. Theor. Biol.* **216**: 147–158.

Kramer, E.M. (2004) PIN and AUX/LAX proteins: their role in auxin accumulation. *Trends Plant Sci.* **9**: 578–582.

Kramer, E.M. (2006) Wood grain pattern formation: A brief review. *J. Plant Growth Regul.* **25**: 290–301.

Kramer, E.M. (2008) Computer models of auxin transport: a review and commentary. *J. Exp. Bot.* **59**: 45–53.

Kramer, E.M. & Bennett, M.J. (2006) Auxin transport: a field in flux. *Trends Plant Sci.* **11**: 382–386.

Kramer, E.M. & Borkowski, M.H. (2004) Wood grain patterns at branch junctions: modeling and implications. *Trees* **18**: 493–500.

Kramer, E.M. & Groves, J.V. (2003) Defect coarsening in a biological system: the vascular cambium of cottonwood trees. *Phys. Rev. E* **67**: 041914.

Lambrecht, M., Okon, Y., Broek, A.V. & Vanderleyden, J. (2000) Indole-3-acetic acid: a reciprocal signalling molecule in bacteria–plant interactions. *Trends Microbiol.* **8**: 298–300.

Leyser, H.M.O. (2006) Dynamic integration of auxin transport and signalling. *Curr. Bio.* **16**: R424-R433.

Li, F., Long, T., Lu, Y., Ouyang, Q. & Tang, C. (2004) The yeast cell-cycle network is robustly designed. *Proc. Natl Acad. Sci. USA.* **101**: 4781–4786.

Lomax, T.L., Mehlhorn, R.J. & Briggs, W.R. (1985) Active auxin uptake by zucchini membrane vesicles: quantitation using ESR volume and DpH determinations. *Proc. Natl Acad. Sci. USA.* **82**: 6541–6545.

Lungley, D.R. (1973) The growth of root systems: a numerical computer simulation model. *Plant & Soil* **38**: 145–159.

Lynch, J.P., Nielsen, K.L., Davis, R.D. & Jablokow, A.G. (1997) SimRoot: Modelling and visualization of root systems. *Plant & Soil* **188**: 139–151.

Malamy, J.E. (2005) Intrinsic and environmental response pathways that regulate root system architecture. *Plant Cell Environ.* **28**: 67–77.

Mitchison, G.J. (1977) Phyllotaxis and the Fibonacci series. *Science* **196**: 270–275.

Mitchison, G.J. (1980a) The dynamics of auxin transport. *Proc. R. Soc. London B. Biol. Sci.* **209**: 489–511.

Mitchison, G.J. (1980b) A model for vein formation in higher plants. *Proc. R. Soc. London B. Biol. Sci.* **207**: 79–109.

Mitchison, G.J. (1981) The polar transport of auxin and vein patterns in plants. *Phil. Trans. R. Soc. London B* **295**: 461–471.

Morohashi, M., Winn, A., Borisuk, M., Bolouri, H., Doyle, J. & Kitano, H. (2002) Robustness as a measure of plausibility in models of biochemical networks. *J. Theoret. Biol.* **216:** 19–30.

Muller, A., Guan, C., Galweiler, L., Tanzler, P., Huijser, P., Marchant, A., Parry, G., Bennett, M.J., Wisman, E. & Palme, K. (1998) AtPIN2 defines a locus of Arabidopsis for root gravitropism control. *EMBO J.* **17:** 6903-6911.

Müssig, C., Shin, G.-H. & Altmann, T. (2003) Brassinosteriods promote root growth in *Arabidopsis*. *Plant Physiol.* **133:** 1261–1271.

Nemhauser, J.L., Hong, F. & Chory, J. (2006) Different plant hormones regulate similar processes through largely nonoverlapping transcriptional responses. *Cell* **126:** 467–475.

Noh, B., Murphy, A.S. & Spalding, E.P. (2001) Multidrug resistance-like genes of *Arabidopsis* required for auxin transport and auxin-mediated development. *Plant Cell* **13:** 2441–2454.

Ohwaki, Y. & Tsurumi, S. (1976) Auxin transport and growth in intact roots of Vicia faba. *Plant Cell Physiol.* **17:** 1329-1342.

Pagès, L. (2000) How to include organ interactions in models of the root system architecture? The concept of endogenous environment. *Ann. Forest Sci.* **57:** 535–541.

Pagès, L. (2002) Modeling root system architecture. In: *Plant Roots: The Hidden Half*, 3rd Edn (eds Y. Waisel, A. Eshel, and U. Kafkafi). Marcel Dekker, New York.

Pagès, L., Vercambre, G., Drouet, J.L., Lecompte, F., Collet, C. & Le Bot, J. (2004) Root Typ: a generic model to depict and analyse the root system architecture. *Plant & Soil* **258:** 103–119.

Paine, P., Moore, L. & Horowitz, S. (1975) Nuclear envelope permeability. *Nature* **254:** 109–114.

Parry, G., Marchant, A., May, S. *et al.* (2001) Quick on the uptake: characterization of a family of plant auxin influx carriers. *J. Plant Growth Reg.* **20:** 217–225.

Petrasek, J., Mravec, J., Bouchard, R. *et al.* (2006) PIN proteins perform a rate-limiting function in cellular auxin efflux. *Science* **312:** 914–918.

Reinhardt, D. (2005) Phyllotaxis - a new chapter in an old tale about beauty and magic numbers. *Curr. Opin. Plant Biol.* **8:** 487-493.

Reinhardt, D., Mandel, T. & Kuhlemeier, C. (2000) Auxin regulates the initiation and radial position of plant organs. *Plant Cell* **12:** 507-518.

Reinhardt, D., Pesce, E., Steiger, P., Mandel, T., Baltensperger, K., Bennett, MJ., Traas, J., Friml, J. & Kuhlemeier, C. (2003) Regulation of phyllotaxis by polar auxin transport. *Nature* **426:** 255-260.

Richter, E. & Ehwald, R. (1983) Apoplastic mobility of sucrose in storage parenchyma of sugar beet. *Physiol. Plant.* **58:** 263–268.

Robinson, D. (1994) The responses of plants to non-uniform supplies of nutrients. *New Phytologist* **127:** 635–674.

Rolland-Lagan, A.-G. & Prusinkiewicz, P. (2005) Reviewing models of auxin canalization in the context of leaf vein pattern formation in *Arabidopsis*. *Plant J.* **44:** 854–865.

Roman, G., Lubarsky, B., Kieber, J.J., Rothenberg, M. & Ecker, J.R. (1995) Genetic analysis of ethylene signal transduction. *Genetics* **139:** 1393.

Runions, A., Fuhrer, M., Lane, B., Federl, P., Rolland-Lagan, A.-G. & Prusinkiewicz, P. (2005) Modeling and visualization of leaf venation patterns. *ACM Trans. Graph.* **24:** 702–711.

Santelia, D., Vincenzetti, V., Azzarello, E., Bovet, L., Fukao, Y., Duchtig, P., Mancuso, S., Martinoia, E. & Geisler, M. (2005) MDR-like ABC transporter AtPGP4 is involved in auxin-mediated lateral root and root hair development. *FEBS Lett.* **579:** 5399–5406.

Shipman, P. & Newell, A. (2005) Polygonal planforms and phyllotaxis in plants. *J. Theo. Bio.* **236:** 154-197.

Smith, R.S., Guyomarc'h, S., Mandel, T., Reinhardt, D., Kuhlemeier, C. & Prusinkiewicz, P. (2006) A plausible model of phyllotaxis. *Proc. Natl Acad. Sci. USA.* **103:** 1301–1306.

Somma, F., Hopmans, J.W. & Clausnitzer, V. (1998) Transient three-dimensional modeling of soil water and solute transport with simultaneous root growth, root water and nutrient uptake. *Plant & Soil* **202:** 281–293.

Stieger, P.A., Reinhardt, D. & Kuhlemeier, C. (2002) The auxin influx carrier is essential for correct leaf positioning. *Plant J.* **32:** 509-517.

Swarup, R., Kramer, E.M., Perry, P., Knox, K., Leyser, H.M.O., Haseloff, J., Beemster, G.T.S., Bhalerao, R. & Bennett, M.J. (2005) Root gravitropism requires lateral root cap and epidermal cells for transport and response to a mobile auxin signal. *Nat. Cell Biol.* **7:** 1057–1065.

Szponarski, W., Guibal, O., Espuna, M., Doumas, P., Rossignol, M. & Gibrat, R. (1999) Reconstitution of an electrogenic auxin transport activity mediated by *Arabidopsis thaliana* plasma membrane proteins. *FEBS Lett.* **446:** 153–156.

Teale, W.D., Paponov, I.A., Palme, K. (2006) Auxin in action: signalling, transport and the control of plant growth and development. *Nat. Rev. Mol. Cell Biol.* **7:** 847-859.

Terasaka, K., Blakeslee, J.J., Titapiwatanakun, B. *et al.* (2005) PGP4, an ATP binding cassette p-glycoprotein, catalyzes auxin transport in *Arabidopsis thaliana* roots. *Plant Cell* **17:** 2922–2939.

Thaler, P. & Pagès, L. (1998) Modelling the influence of assimilate availability on root growth and architecture. *Plant & Soil* **201:** 307–320.

Thornley, J.H.M. & Johnson, I.R. (2000) *Plant & Crop Modelling.* Blackburn Press, Caldwell, NJ.

Tsurumi, S. & Ohwaki, Y. (1978) Transport of 14C-labeled indoleacetic acid in Vicia root segments. *Plant Cell Physiol.* **19:** 1195-1206.

Walch Liu, P., Ivanov, I.I., Filleur, S., Gan, Y.B., Remans, T. & Forde, B.G. (2006) Nitrogen regulation of root branching. *Ann. Bot.* **97:** 875–881.

Walk, T.C., Jaramillo, R. & Lynch, J.P. (2006) Architectural tradeoffs between adventitious and basal roots for phosphorus acquisition. *Plant & Soil* **279:** 347–366.

Wierzbicka, M. (1987) Lead accumulation and its translocation barriers in roots of *Allium cepa* L. – autoradiographic and ultrastructural studies. *Plant Cell Environ.* **10:** 17–26.

Yang, Y., Hammes, U., Taylor, C., Schachtman, D.P. & Nielsen, E. (2006) High-affinity auxin transport by the AUX1 influx carrier protein. *Curr. Biol.* **16:** 1123–1127.

Index

absolute proteome quantification 52–53
abundance (*see* transcript abundance)
access to archived data 1–2
acetylation, post-translational 145
action potentials 176–177, 179–180, 182
activator function 116
adaptation, chemotaxis-related 165
affinity tags 49–52
alcohol dehydrogenase 82
alcohol production 80–88
allosteric regulation 73
alternate splicing 37–38
amino acid pools 86
amplification of nucleic acids 17–33, 147–149
anatomical mammalian heart models 184–186
animal models iii, 123–124, 184
apical meristems 203–204
apoplastic diffusion 201–202
aptamers 153
AQUA synthesis 52
Arabidopsis thaliana:
    cell fate determination 115
    flower development 120, 124
    gene co-expression microarray data 4–8
    leaf and root hair pattern formation 124–125
    molecular basis of signal integration 206
    primary root system simulation 197, 204
archiving standards 1–2
aRNA 24, 26
array-cgh (*see* DNA-based comparative genomic hybridization on arrays)
At5g59690 *Arabidopsis* histone H4 gene 4–7
ATP production 66
atrioventricular node 177
attractants, chemotaxis 161–172
autophosphorylation 165
autoregulatory loops 103–108
autoregulatory transcription factors 97–98
AUX1 auxin influx carrier protein iv, 202–203
auxin transport iv, 195, 198–206

Bacillus subtilis 127–135
bacterial chemotaxis 161–172
Bayesian Dirichlet (BDe) scoring metric 130
Bayesian networks method 13, 14
Bayesian Structure Learning 129–135
B cells 14
BDe (*see* Bayesian Dirichlet scoring metric)
Beeler–Reuter cardiac ventricular model 180
biofilms 166
bioinformatic approaches 1–15
b ions 40
bistability 142, 147–149
body plan development 126–127
Boolean networks 118–121
bottom-up approaches 175
boundary conditions 169, 171
brassinosteroids 206
Brownian motion 117

Ca^{2+} ions 175–177, 180–181
cable equation 182
CAD (*see* collisionally activated dissociation)
calcium ions 175–177, 180–181
calmodulin-binding protein At2g43040 8
cancer 37, 99–100, 142
carbon-13 glucose 49
cardiac (*see* heart)
carriers, auxin transport 201
cell differentiation 114–115, 117–118
cell fate determination 115, 120
CellML XML-based encoding language 186–187
cell sampling 17–31
cellular characterization 17–18, 153
cellular composition 20
cellular interaction networks 96–98
cellular models 175, 179–180, 199
cgh (*see* DNA-based comparative genomic hybridization on arrays)
chemotaxis, bacterial i–ii, 161–172

Che proteins i–ii, 163–172
ChIP analyses 6–7
chromatin 116
circular regulation coefficient 70
codes for numerical solutions 186–187
co-expression analysis 4–9, 12
collisionally activated dissociation (CAD) 40, 43, 45
co-mentions in literature searches 10
commercial tools 27, 42–43, 171
commotio cordis 178
comparative genomic hybridization (cgh)
      (*see* DNA-based comparative genomic
      hybridization on arrays)
complexity of the proteome 37–38
computational models (*see* mathematical models)
COMSOL computational packages 171
conductivity tensors 182
connectivity between ions 40
constrained inferences 99–100
continuum mathematical theory of reaction-
      diffusion equations 167–171
continuum models 199
controlled vocabularies 1–2, 10
coupled electromechanical models 185–186
cysteine residues 48

database access 1–2
data generation 11–15
data integration concept 1
data retrieval, public 2–11
data search terms 10
DE (*see* differential equation models)
degeneracy 132
degenerate oligonucleotide primed PCR (DOP) 30
DELLA proteins 206
deltaKPQ mutation 187–188
demethylation 163
Depth First Search (DFS) algorithm 132
deterministic Boolean networks 118–121
developmental biology:
      constraints, epistasis 121–123
      Drosophila melanogaster embryo segmentation
         94, 101–102
      gene regulatory network models 113–135
      plant root growth 195–206
DFS (*see Depth First Search* algorithm)
difference gel electrophoresis (DiGE) 48
differential equation (DE) models 96–98
diffusion 117, 150–152, 198–200, 201–202
DiFrancesco–Noble model of Purkinje cells 180
DiGE (*see* difference gel electrophoresis)
dimensional parameters 167, 171
discretization 198–199
distribution-moment models 181
DNA:
      (*see also* gene…; genome…)
      hierarchical subdivision of cellular
         biochemistry 65–66
      microarrays 4–7, 17–18
      supercoiling, *Escherichia coli* 70

domains 142–144
DOP (*see* degenerate oligonucleotide primed
      PCR)
Drosophila melanogaster 94, 101–102, 123–124
dynamic expression data 93–108
dynamic gene regulatory network models 113–135
dynamics, spatio–temporal 141–153

Eberwine protocol 23, 26, 31
ECD (*see* electron capture dissociation)
ECG (*see* electrocardiography)
efflux auxin carriers iv, 202, 204, 298
EGF (*see* epidermal growth factor)
electrical–mechanical coupling 177, 185–186
electrical models 179–180, 182–183, 185–186
electrocardiography (ECG) 177
electron capture dissociation (ECD) 43, 47
electron transfer dissociation (ETD) 43, 45, 47
elongation root zone iv, 202–203
embryo segmentation 94, 101–102, 123, 124
endosomes 151–152
enhancers 116
ensembles of surrogates 134
environmental robustness 122–123
epidermal growth factor (EGF) 142
epidermal hair cells 124–125
epigenetic landscape 121
epistasis 113, 121–123
epitope-tagged proteins 7
Equivalent Sample Size (ESS) 131
Escherichia coli:
      chemotaxis and intracellular signalling i–ii,
         161–172
      DNA supercoiling 70
      gene regulatory networks 123
      lac operon 66
      metabolic system reconstruction 4
      two-colour spotted microarrays 129
ESS (*see* Equivalent Sample Size)
ETD (*see* electron transfer dissociation)
ethylene 206
evolution 113–114, 126–127
exponential strategies 24–26, 28–33
expression (*see* gene expression)

fading-memory models 182
false positives 8, 46
feedback control 66, 70, 146–149, 178
fermentation 80–88
Fibonacci series 203–204
fibrillation models iii, 188–189
Fick's law 199–200
finite element methods 183, 184, 186
flagellar motion 161–172
FliM protein motors 161, 163, 166, 172
flow cytometry studies 14
flower development 120, 124
fluorophores 48
flux boundary conditions 169
flux control coefficients 69

FSPM (*see* functional structural plant models)
functional association prediction systems 11
functional interaction network forms 94
functional–structural models 175, 195–206
functional structural plant models (FSPM) 196–197
functions choice, differential equations 97

GA (*see* gibberellins)
gain 161–162, 165
GAP (*see* GTP-hydrolysis activating protein)
gap gene networks 101–103
gas-phase peptide ion fragmentation 40–44
GEF (*see* guanine nucleotide exchange factor)
gel electrophoresis-based approaches 38, 39
gene expression:
   Arabidopsis histone genes 4–7
   assays 17–18
   co-expression analysis 4–9
   Hierarchical Regulation Analysis 75, 77
   mathematical model data 93–108
   metabolic regulation distinction 65
   regulatory networks 12, 113–135
   RNA level analysis 18–21
gene network inference methods 128–135
Gene Ontology (GO) 2
gene polymorphisms 187–188
gene promoters 5–7
gene regulatory network (GRN) models 12, 113–135
Genevestigator Metanalyzer™ 8–9
genome level gene regulatory network models 127–135
genome sequence annotation 1–2
gibberellins (GA) 206
global nucleic acid amplification 17–33
GLT (*see* glucose transporter)
glucose phosphorylation 81
glucose transporter (GLT) 85–88
glycolysis, yeast 65, 66, 79, 80–88
GO (*see* Gene Ontology)
gradients, spatial intracellular 141, 150–152
gravitropism 202–203
greedy searches 132
GRN (*see* gene regulatory network)
growth factor signalling 142–144
GTP-hydrolysis activating protein (GAP) 145
guanine nucleotide exchange factor (GEF) 145

hair cells 124–125
HCA (*see* Hierarchical Control Analysis)
heart:
   fibrillation models iii, 188–189
   ion channel phenotypes of gene polymorphisms 187–188
   mammalian iii, 175–189
   physiology 176–179
   structure/function 176–177
   tissue models 182–184
   whole organ models 184–186
helix-loop-helix transcription factors 104–107

hepatocyte growth factor (HGF) 142
heritable perturbations 122
Hes1 murine transcription factor 103–108
HGF (*see* hepatocyte growth factor)
Hierarchical Control Analysis (HCA) 69
hierarchical regulation 65–90
Hierarchical Regulation Analysis (HRA) 70–88
high-amplitude oscillations 107
high-throughput microarray studies 9–10, 22–30
histone H4 gene 4–7
Hodgkin–Huxley cardiac cell model 179, 182
HRA (*see* Hierarchical Regulation Analysis)
human immune cell signalling 14
HUPO Proteome Standards Intitiative 43, 46
hybrid networks 3
hydroxylation, post-translational 145
hysteresis 149

immobilized metal ion affinity chromatography (IMAC) 47
immune cell signalling 14
incompressibility, myocardium 183
inferences:
   gene regulatory networks 113–114, 128–135
   Hes1 murine transcription factor expression 107
   partial network data 101–108
   transcription factor expression profiles 99–100
initial diffusion conditions 169, 171
insects 94, 101–102, 123–124, 127
instantaneous rate of change 96–98
integrative systems biology (ISB) 195, 197–198, 202–203
intracellular gradients 141, 150–152, 201
intracellular signalling 3, 14, 141–153, 162–172
ion channel phenotypes 187–188
ISB (*see* integrative systems biology)
isobaric tags for relative and absolute quantification (iTRAQ™) 50–51
isotope labelling 48–53
iTRAQ™ (*see* isobaric tags for relative and absolute quantification)

kinases 141–153
kinetics, proteome 53–54

L1 layer cell model 205–206
labelling 48–52
lac operon 66
landscape metaphor 121
Langevin equation 117
languages, controlled 1–2, 10
law of mass action 168
length scales 170
linear RNA amplification protocol 23, 31–33
literature analysis 10
logarithms 79, 130
long QT syndrome (LQTS) 187
low duty MS cycle 53
lower limit condition 28–30

LQTS (*see* long QT syndrome)
Luo–Rudy model 180

macro-variables 143–144
MAGIC system 11
magnetic resonance imaging (MRI) 184
MALDI TOF MS (*see* matrix-assisted laser desorption/ionization time-of-flight mass spectrometry)
maleimide-linked dyes 48
mammalian heart modelling 175–189
MAPK (*see* mitogen-activated protein kinase)
Markov Chain Monte Carlo (MCMC) methods 13
mass action, law of 168
massively parallel signature sequencing (MPSS) 22
mass spectrometry (MS) 39–46, 53
mathematical models:
    bacterial intracellular signalling 163–172
    COMSOL 171
    dynamic expression data 93–108
    growth factor signalling 142–144
    hierarchical regulation 65–90
    interaction networks 95
    mammalian heart 175–189
    partial networks 103–108
    plant root growth development 195–206
    time-course expression data 98–102
matrix-assisted laser desorption/ionization time-of-flight mass spectrometry (MALDI TOF MS) 40, 41
MCA (*see* Metabolic Control Analysis)
MCMC (*see* Markov Chain Monte Carlo)
MCP (*see* methyl-accepting chemotaxis proteins)
mechanical models, cardiac 180–186
mechanistic models 95–96, 143–144
mechano–electrical feedback 178
membrane-bound enzymes 150–152
meristem 203–204
Metabolic Control Analysis (MCA) 67, 69
metabolic networks 2–11, 12
metabolic regulation 65, 67–70, 80
    coefficient 73, 75, 83–84
metabolic stable isotope incorporation into proteins 48–49
methyl-accepting chemotaxis proteins (MCP) 161–172
methylation 145, 163, 165
O-methyl isourea 51
MI (*see* Mutual Information)
MIAME (Minimum Information About a Microarray Experiment) 1
Michaelis–Menten kinetics 72, 148
microarrays 1, 4–7, 17–18, 127–135
micro-states of a network 143
microtubules 151–152
Minimum Information About a Microarray Experiment (MIAME) 1
minute mRNA samples 28–30
mitogen-activated protein kinase (MAPK) 142

modelling:
    (*see also individual types of modelling*)
    bacterial chemotaxis 161–172
    mammalian heart iii, 175–189
    plant root development 195–206
model reduction 166
molecular components lists 94
molecular motor-mediated movement 151–152, 161, 163, 166, 172
monitoring techniques:
    cellular proteins 153
    gel electrophoresis 38, 39
    isotope labelling 48–53
    magnetic resonance imaging 184
    mass spectrometry 39–46, 53
    microarrays 1, 4–7, 17–18, 127–135
    MPSS 22
    peptide mass fingerprinting 40–46
    shotgun proteomics 38, 39, 46
    Western blots 153
morphogenesis:
    (*see also* developmental biology)
    Arabidopsis cell fate determination 115
    pattern formation 114–115, 117–118
    plant root growth modelling 196–206
motifs, gene regulatory 125–126
movement, bacterial flagellae 161–172
MPSS (*see* massively parallel signature sequencing)
MRI (*see* magnetic resonance imaging)
mRNA 22–33, 65–66, 75–76, 78, 99–100
MS/MS (*see* tandem mass spectrometry)
multi-core CPUs 187
multidimensional peptide chromatography (*see* shotgun proteomics)
multiple independent search programs 46
multiple parallel reactions 143–144
multiple receptor states 143–144
multiple re-entrant waves 188
multiple transcription factors 101–102
multiple translated proteins 37–38
multi-scale models 166, 172, 195
murine transcription factor Hes1 103–108
Mutual Information (MI) 127–128, 133–135
MYC transcription factor 14
myocardium, mammalian 182–184

Na^+ ion channels 175–176, 188
negative autoregulation 104
negative feedback 146–149
nerve growth factor (NGF) 142
network prediction difficulties 12–13
Newton's laws of motion 117, 183
NGF (*see* nerve growth factor)
nitrogen-15 ammonium chloride 49
nitrogen starvation 80–88
noise 96, 152–153
non-coding RNA 19
non-dimensional parameters 167, 170–171

non-heritable transient perturbations 122
nucleic acid amplification 17–33
numerical aspects of models 186–187

oncogenes 142
one-dimensional bifurcation diagrams 149
one-dimensional gel electrophoresis 38
Op18/stathmin 151
organ scale modelling 175
oscillations 104–107, 147–148

p53 tumour suppressor transcription factor
    99–100
parameter selection/validation 103
partial differential equations (PDE) 118, 167–171
partial network data inferences 101–108
pattern formation 114–115, 117–118, 124–125
PC12 cell-line 142
PDE (see partial differential equations)
PEP (see primer extension preamplification)
peptide mass fingerprinting (PMF) 40–46
perturbations of networks 14, 74, 122
PH (see pleckstrin homology)
phenotypes 121–123
phosphatases 141–153
phosphoroamidate chemistry 47
O-phosphorylation 38
phosphorylation 38, 46–47, 53–54, 145
phosphotransfer 163–171
phosphotyrosine binding (PTB) domain 143
phyllotaxis 195, 198–206
physiological modelling 175–189
physiology, cardiac 176–179
PIN1 auxin efflux carrier protein iv, 204
PIN2 auxin efflux carrier protein iv, 198, 202
plants iv, 123–126, 195–206
pleckstrin homology (PH) domain 143
PMF (see peptide mass fingerprinting)
pollen-specific calmodulin-binding protein
    At2g43040 8
poly-(A) tailed transcripts 19
polymerase chain reaction (PCR) 24–26, 28–33
pooling of samples 22
populations of cells 166, 172
positive feedback 146–149
post-synthetic stable isotope incorporation 49–52
post-transcriptional regulation 116
post-translational modifications (PTM) 37, 38,
    46–48, 144–149
primer extension preamplification (PEP) 30
probability distributions 13
process definition in HCA 79
promoter regions 116
propagating waves 151–152
proportional regulation paradigm 88–89
proteases 78
proteins:
    amino acid pools 86
    Che proteins i–ii, 163–172

hierarchical subdivision of cellular
    biochemistry 65–66
interaction network concepts 12
level regulation 77
modification 37, 38, 46–48, 141–153
phosphorylation 151–152
profiles 99–100
proteome analysis 37–54
recognition 39–46
PTB (see phosphotyrosine binding)
PTM (see post-translational modifications)
public data sources 2–11, 127
Purkinje cells 180
pyruvate 86

QconCAT quantification method 52
qualitative proteomics methods 38–48
quality of cell samples 21
quantitative studies 37–38, 48–53, 94–102, 147
quantity requirements 22–33

rabbit heart iii, 184
Raf-1 activation 141–142
random perturbations 117
random-walk-like behaviour 161, 163
reaction-diffusion equations 167–171
receptor states 143–144
receptor tyrosine kinases (RTK) 142–143
regulation analysis (see Hierarchical Regulation
    Analysis)
regulatory network models 113–135
relative proteome quantification 48–52
repression threshold 105–107
repressor function 116
reverberation coefficients 70
reverse engineering 12
reverse transcriptase 23, 25
Rhodobacter sphaeroides 161, 172
RNA:
    amplification 22–33
    gene expression 18–21
    Hierarchical Regulation Analysis 75–76, 78
    hierarchical subdivision of cellular
        biochemistry 65–66
    transcription factor expression 99–100
robustness 113–114, 121–123, 202
root system models iv, 195–206
RTK (see receptor tyrosine kinase)

Saccharomyces cerevisiae (see yeast)
SAGE high-throughput technique 19, 22
SAM (see shoot apical meristem)
sample amplification 22
sample size 21
sarcoplasmic reticulum (SR) 177, 180–181
scales of modelling 175
SCN5A gene 187–188
search terms 10
search tools 42–43

segment polarity, insects 127
sensitivity:
  bacterial chemotaxis 163, 165
  Hes1 murine transcription factor expression
    105–107
  protein modification cycles 145, 147
sequence databases 43
sequestration effect 145
shoot apical meristem (SAM) 203–204
shotgun proteomics 38, 39, 46
signal amplification 22
signal transduction 3, 14, 141–153, 161–172
SILAC (*see* stable isotope labelling with essential
    amino acids in cultured cells)
simplified modelling methods 196
simulation (*see* modelling)
Singular Value Decomposition 128
sinoatrial node 177
size of RNA samples 28–30
sliding filament theory 181
small sample size 28–30
SMART ᵀᴹ PCR technology 25
Smith's phyllotaxis model 205
software 42–43, 125, 171
source codes, numerical solutions 186–187
spatial aspects 94, 150–152, 198
spatio–temporal aspects 116–117, 141–153,
    161–172
species data 10
spiral phyllotaxis 203–204
squid giant axon 179–180, 182
SR (*see* sarcoplasmic reticulum)
$S(t)$ (*see* state of the network)
stable isotope labelling 48–53
stable isotope labelling with essential amino acids
    in cultured cells (SILAC) 49
standards 1–2, 43, 46
state evolution models 95–96
state of the network ($S(t)$) 95
stathmin-oncoprotein 18 (Op18/stathmin) 151
steady-state values 88
stochastic effects 29–30
STRING system 11
structural aspects 94–95, 133
structure learning method 13
surrogates ensembles, Mutual Information 134
switches (*see* bistable dynamics)
systems identification 3

T7 RNA polymerase 23–25
tags, stable isotopes 49–52
tandem mass spectrometry (MS/MS) 39, 40–46
template switching 23, 25

timescales:
  cell sample harvesting 20
  Escherichia coli phosphotransfer reaction-
    diffusion equations 170
  expression data, mathematical models 98–102
  hierarchical regulation analysis 65–90
  RNA metabolism 75–76
  temporal dynamics 144–149
tissue level scale 175, 177–178, 182–184
titanium dioxide beads 47
top-down approaches 175
topologies of interaction networks 94
transcript abundance 19, 28
transcription-controlled fluxes 67–70, 80
transcription factors 97–100
  (*see also individual transcription factors*)
transcriptome analysis 18–21
translational regulation 116
translational start sites 37–38
translation rate constant 76
transport, auxin iv, 195, 198–206
Trypanosoma brucei flagella 41, 50–51
tryptic peptide sequence selection 53
tumbling motion, bacterial chemotaxis 161
tumour suppressor transcription factor p53 99–100
two-colour spotted microarrays 128–129
two-dimensional gel electrophoresis 38–39

ubiquitylation, post-translational 145
UCSD Continuity package 187
ultrasensitivity 145, 147
updating regimes 121

validation 10, 46, 103
ventricles, cardiac iii, 179–180, 184, 188–189
vertical systems (*see* hierarchical regulation)

waves of protein phosphorylation 151–152
Western blots 153
whole-genome amplification (WGA) 30
whole heart models 184–186
workflows 38–39

yeast:
  alcohol production after nitrogen starvation
    80–88
  cell cycle-related protein complex formation 11
  glycolysis 65, 79, 85–89
  2-hybrid protein interactions, high-throughput
    data 9–10
  metabolic system reconstruction 4
  transcription factors 6–7
y ions 40, 45

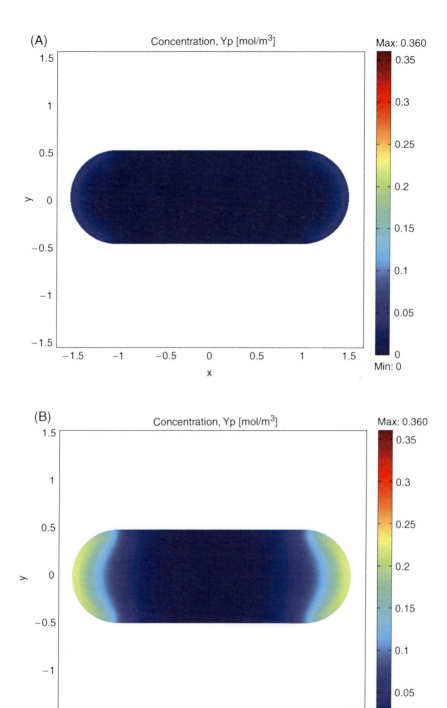

Chapter 8, Figure 3. The dynamic change in the non-dimensional concentration of CheY$_P$ within an E. coli cell at (*A*) 2 ms, (*B*) 10 ms, (*C*) 20 ms, and (*D*) 40 ms. Here the cell has been taken to be 3 μm long and 1 μm wide.

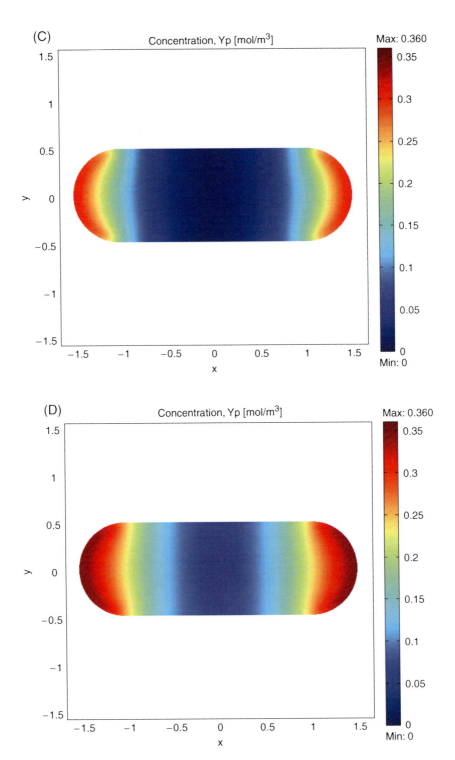

Chapter 8, Figure 3. cont'd.

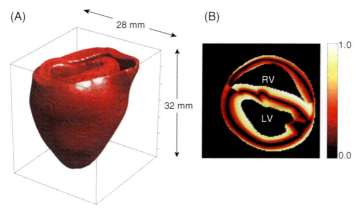

Chapter 9, Figure 2. *Anatomically detailed model of the rabbit ventricles. (A) Isosurface view of Cartesian geometry, showing dimensions of bounding box. (B) Short axis slice at mid height, showing left and right ventricular cavities (LV and RV), and with colour coding showing the direction cosine of fibres relative to the image plane. Dark colours show circumferentially oriented fibres in the plane of the image, and light colours show fibres oriented orthogonal to the image plane.*

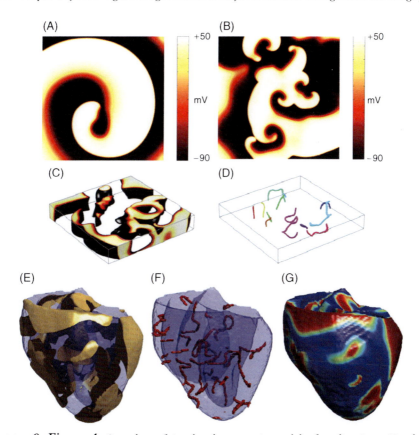

Chapter 9, Figure 4. *Snapshots of simulated re-entry in models of cardiac tissue. Simulations of single (A) and multiple (B) re-entrant waves in 2-D tissue sheet with dimensions 12 × 12 cm. (C) Simulation of multiple re-entrant waves in a 3-D tissue slab with rotational anisotropy. (D) Filaments corresponding to the snapshot in (C). (E–F) Simulation of multiple re-entrant waves in anatomically detailed model of the canine heart; (E) Isosurface view, where yellow regions enclose active tissue; (F) Filaments corresponding to (E). (G) Surface activation corresponding to (E), where red regions are active tissue.*

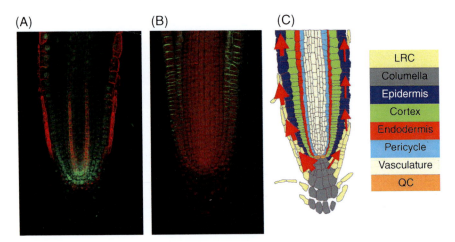

Chapter 10, Figure 2. *The direction of auxin transport is determined by the asymmetric localization of AUX1 and PIN auxin carriers. (A) Confocal image of root apical tissues (in green) expressing the auxin influx carrier AUX1 (red); (B) Confocal image of root apical tissues (in red) expressing the auxin efflux carrier PIN2 (in green); (C) Schematic of root apical tissues illustrating the asymmetric redistribution and direction of auxin flux following a gravity stimulus (denoted by red arrows). Root tissues are colour coded according to the inset key.*

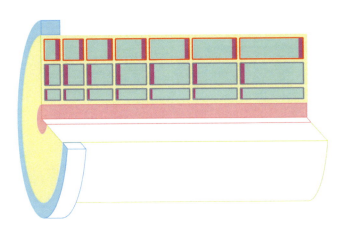

Chapter 10, Figure 3. *Schematic illustration of the computer model of the root elongation zone used by Swarup et al. (2005). Cells are represented as green rectangles and the apoplast is in yellow. The model includes the three outermost cell layers: the epidermis, cortex, and endodermis (top to bottom). Auxin is deposited into the apoplast of the distal elongation zone from the lateral root cap (blue region), and auxin that reaches the stele (pink region) is trapped and removed from the model system. The model includes the effects of auxin diffusion in the apoplast and in the cytoplasm of cells, and it includes the spatial distribution and activity of influx and efflux carriers (orange and plum respectively). For cell membranes without influx carriers (grey), auxin can only enter the cell by diffusive permeation of the membrane.*